Theory and New Applications of Swarm Intelligence

Theory and New Applications of Swarm Intelligence

Editor

Bilroy Muller

scitus
academics

Theory and New Applications of Swarm Intelligence
Edited by **Bilroy Muller**

ISBN: 978-1-68117-586-7
Library of Congress Control Number: 2016946441

SCITUS Academics LLC,
616, Corporate Way, Suite 2, 4766,
Valley Cottage, NY 10989

www.scitusacademics.com

Printed in United States of America on Acid Free Paper

Preface

A single ant or bee is not as smart as its colonies. Swarm intelligence studies provide insights that help people manage complex systems, from lorry routing to military robots. A colony can solve unthinkable problems for individual ants, such as finding the fastest way to the best food source, assigning workers to different tasks or possibly defending a territory from neighbours. Ants may be small dummies as individuals, but they respond efficiently and rapidly to their environment as colonies. They do it with so-called swarm intelligence. Swarm Intelligence is a discipline dealing with artificial and natural systems made up of many people who coordinate by means of decentralized control and self-organization. In particular, the discipline focuses on the collective behavior that stems from the local interactions between people and their environment. Swarm Intelligence has a distinct multidisciplinary character, as systems with the above characteristics can be found in a range of fields. In general, SI systems consist of a population of simple agents that interact locally with each other and with their environment. Normally, inspiration comes from nature, especially from natural systems. The agents follow simple rules, and although there is no centralized control structure dictating how individual agents should behave, area and to a certain degree randomly, interactions between these agents lead to the development of "intelligent" global behavior, unknown to the agents themselves.

This state of the art coverage enables the reader to know more both theoretical and technical aspects and Swarm Intelligence applications that reflect the emerging trends in algorithms. This book therefore presents some recent advances in Swarm Intelligence, in particular in new swarm-based optimization methods and hybrid algorithms for a number of uses.

Table of Contents

Chapter 1

An Open Flow Controller Based on the Intercerebral Neural Network for Media Independent Handover

Qiong Wu, Jun S Huang* and Oliver WW Yang

3rd Department, Suqian College, Jiangsu 223800, P.R. China

ABSTRACT

This paper presents a novel open flow controller based on the intercerebral neural network for the media independent handover over wireless operation. This controller makes a binary decision based on the link control parameters obtained instantaneously from the trained neural network, which in turn depends on the instantaneous mobile speed interaction with the wireless link performance parameters. These parameters of packet loss, round trip time, radio signal strength, bandwidth, data cost and vehicle speed are typically different from network to network. We use a mutated particle swarm optimization to train the fundamental controller equation for the media independent handover that are mixed with the radial and sigmoid activation functions of the intercerebral neural network. Simulations based on the field experiment test data show that the system using the swarm intelligence algorithm is practical for the improvement of the overall secure networking experience for the fast delivery.

KEYWORDS

Handover; Neural; Radial; Sigmoid; Swarm

1. INTRODUCTION

A transportation company for fast delivery service usually has to maintain a team of service trucks on the road, each of which is equipped with a mobile scanner that keeps track of the goods during its upload and download operations as well as reporting the GPS position of the truck. The dispatching schedule and the route of each truck are semi-fixed because the route is not fixed like a bus but also not purely random like a tax driver's either. For each truck driver, the schedule and route are dictated by a handful of fixed customers, which are typically optimized by the so-called swan goose algorithm [1] to save the total amount of gasoline needed for the day. The challenges remain to maximize the throughput of the uploading the logistical information, while the dispatching truck is going through the semi-fixed route (relatively fixed route on the roughly fixed schedule).

Currently either the WiFi-based mobile scanner [2] or the GPRS-based mobile scanner are used to support the communication of logistical information which includes pictures of 1-D bar codes or 2-D QR codes, as well as sensitive credit card information, where no breach of information is allowed. WiFi-based operation is fast, but only available within limited areas. Although GPRS-based is more accessible, it is fairly expensive, especially in dealing with large amount of bulky pictures. There is also a security problem when a cheap broken link may allow the attacker to easily obtain information on payment transaction. To minimize this risk; we need a solid and smart algorithm to continuously monitor the overall quality of each network. To handle the economical dilemma, we also need to further improve the performance of the system of scanners by combining the access methods of WiFi and GPRS. We would like to design a smart mobile handover algorithm using the latest evolutional computation algorithm to maximize the secured throughput but with the minimum handover cost between the WiFi and the GPRS media. To achieve this, we look to the implementation using SDN (Software Defined Network).

Started as an approach to computer networking [3] a few years ago, SDN allows network administrators to manage network services through the abstraction of lower level functionalities. A communication system in general has two planes: a control plane that can make decision where traffic is sent, and the data plane that can forward traffic to the selected destination. SDN decouples the control plane in a system from the data plane in the underlying systems. In addition to the simplification of networking issues, we believe it can also lower the overall cost as to be shown in this paper.

There are different communication methods between the control plane and the data plane. We would like to use the Open Flow method from the Open Networking Foundation (ONF) [4], which also markets the use of “cloud computing” before it became popular. Dynamic resource allocation is one important aspect of an elastic Cloud architecture. As the usage of mobile computer operating systems grows, such as the growing number of scanners used in the booming dispatching industry, the need has arisen for an additional layer in SDN. This layer should allow network operators to specify network services (e.g., uploading or downloading express dispatching information and sensitive credit card code) without coupling these specifications with network interfaces (e.g. interfaces to WiFi, WiMAX, GPRS or LTE). This enables entities to move between interfaces without changing identities or violating secured service specifications. It can also simplify network operations, where global definitions per identity do not have to be matched to each and every interface location when a service truck is driving through the entire city or among cities. Such a layer can also reduce some complexity built up in network elements; this is achieved by decoupling the identity and flow-specific control logic from the basic topology-based operations of forwarding, bridging, and routing on top of the wireless handover.

The global software defined control also tracks down specific flow contexts based on source and destination identity aspects. A mechanism for driving network hardware has been added and adopted by network equipment manufacturers for the purpose of sharing edge driving between software-defined edge and vendor-specific bridging and routing. A set of open commands for forwarding was defined in the form of a protocol known as Open Flow. The Open Flow protocol enables globally-aware software controllers, which may be centralized or distributed, to drive the network edge hardware in order to create an easily programmable identity-based overlay on top of the traditional IP core.

The wireless use case enables Open Flow to effectively perform both break-before-make and make-before-break flow assignments by using Media Independent Handover (MIH) protocol commands [5] as specified by IEEE802.21, which is under development within IEEE802 group. It is expected that 802.21 is well suited for Open Flow implementation, because Open Flow is specified in a media independent manner.

Our study further employs the particle swarm training algorithm to optimize the media independent solution. This can be extended to more new networks, such as LTE or WiMAX. There was a demonstration of the previous application of PSO for Computer Numerically Controlled (CNC) machine tool design [6,7], which inspired our new finding that is driven by our industrial sponsor Jufeng Group in Danyang Development Zone, who is dedicated to transportation and storage logistics.

2. NEED FOR AN INTERCELEBRAL NEURAL NETWORK FOR MOBILE HANDOVER CONTROLLER

To minimize the data cost consumed during the dispatching operation, both the WiFi system and GPRS system have to be seamlessly handover; such that the summation of both providing the maximum and secure throughput. As the truck starts or stops there will be a WiFi connection to the office, as the truck moves on the road, there will be GPRS service along the road. Note that during the normal driving period, the GPS data is small, while during the start or stop time, where picture and credit card signature are collected, the data is relatively large and sensitive; we thus need the IEEE802.21 compatible Open Flow controller to switch between the two networks.

Typical wireless handover is determined by looking at the Link Going Down (LGD) parameter [8], LGD can be a function of a number of measured parameters, such as packet loss, round trip time, radio signal strength, bandwidth, data cost and vehicle speed, etc. Each of these parameters has a relationship to the location environment relative to the base station or access point. Take the most used parameter, Radio Signal Strength Indicator (RSSI) as an example, the relationship is not linear with respect to the transmitter to receiver distance; and the typical RSSI value for WiFi and GPRS are also different. As such, when we trying to combine all parameters into one decision parameter, it becomes highly complicated nonlinear, and slightly random, as the schedule and route can be deviated from the daily routing, the best way to learn this kind of the function is through special class of the neural network, which we call it Intercerebral Neural Network (INN).

An intercerebral neuron network structure is proposed instead of one pure kind, for the following reason: Each measured parameters has its own underline nonlinear functions, which are stemmed from completely different roots, by using the exact the same neuron to describe them is hard. Note that the switch from WiFi to GPRS will be different from switch GPRS to WiFi at least. A typical dispatching system is shown in Figure 1. The car A is driving from GPRS 4+1 WiFi into 802.11b, and car B is driving from 802.11b to WiFi 802.11n, car C is from WiFi 802.11n to GPRS 3+2, so on so forth. We need to train the neuron in such a way, that the decision is optimum for all the potential cases.

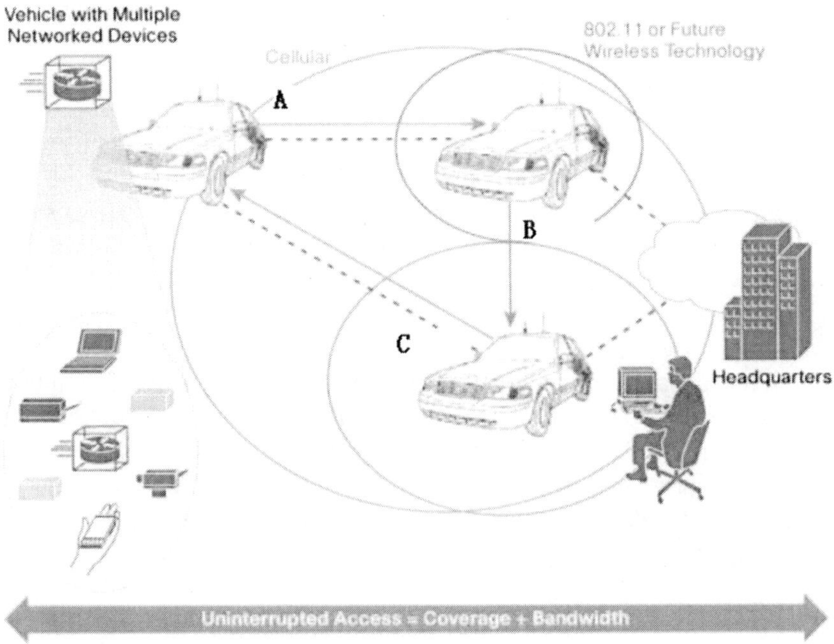

Figure 1: Wireless handover model.

It is important to note that the controller makes the decision by using a Particle Swarm Optimization (PSO) [9] trained intercerebral neural network on the current measured parameters. By monitoring the current parameters, we can automatically balancing the cost with trained system. The algorithm learns how to increase the secured throughput without causing too much extra charges, based on the collected fields' data.

Due to the semi fixed schedule and route, there is the need to enhance previous MIH methods [10] for the following reasons:

The direct calculation model cannot follow the route variation from the extra customer calls occurred on the road.

- The prediction model can only deal with pure random events, but has not been optimized for the relatively fixed route.

- The complicated adaptive model method can only partially deal with the route variation, but does not treat the schedule changes very well.

- The homogeneous Neural Network (NN) can work, but we found that the parameters trained for one truck does not fit the other.

The method that can handle the above four considerations is the INN trained with a mutated PSO method [11]. The intercerebral approach comes from studying of actual neuron structure between man and woman, where we found that there are only two types of neurons in our brain; we call it either RBF [12,13] or SBF:

a) A Radial Basis Function (RBF) network [14] is a neural network that uses radial basis functions as activation functions. The output of the network is a linear combination of radial basis functions of the inputs and neuron parameters. Radial basis function networks have many uses, including function approximation, time series prediction, classification, and system control.

b) A Sigmoid Basis Function (SBF) is a mathematical function having an "S" shape (sigmoid curve). Often, sigmoid function refers to the special case of the logistic function. Another example is the Gompertz curve. It is used in modeling systems that saturate at large values of input, e.g. the ogee curve as used in the spillway of some dams [15,16]. A wide variety of sigmoid functions have been used as the activation function of neurons, including the logistic and hyperbolic tangent functions. Sigmoid curves are also common in statistics as cumulative distribution functions, such as the integrals of the logistic distribution, the normal distribution, and Student's t probability density functions.

c) According to the biological concept, RBF is related to brain memory function, more on our left brain, and SBF is related to brain imagination capability, more on right brain. The structure of RBF supports the academic school of connectivitilism [17], while that of SBF favors the computational believers. In our opinion, SBF offers nonlinear effects for large input value; RBF provides nonlinear effect at small input value. A combination of both SBF and RBF will result in more nonlinear blending across the entire region. We use the mix of both to learn the input and output relations with multiple nonlinear root causes.

d) Psychological studying shown that woman can learn and adapt to new social environment faster than man. A recent study used diffusion MRI to map functional differences [18] in male and female brains. In the cerebral cortex it was observed that, in comparison with the opposite sex, there was greater intracelebral neural communication in male brains and greater intercerebral neural communication in female brains. In another word, woman may have more blending of nonlinear functions in the intercelebrum for them to plasticity the neurons to new culture environment faster. This finding leads us to pay attention to the intercerebral neuron network. The training of such mixed network becomes a bit challenging, for this reason, we picked the mixed PSO method with part of Genetic Algorithm (GA).

PSO is a computational method that optimizes a problem by iteratively trying to improve a candidate solution with regard to a given measure of quality. PSO optimizes a problem by having a population of candidate solutions, here dubbed particles, and moving these particles around in the search-space according to simple mathematical formulae over the particle's position and velocity. Each particle represents a combination of network conditions, say Monday to Friday, good or bad, GPRS or WiFi, that's 5x2x2=20 in total. Each particle's movement is influenced by its local best known position, but is also guided toward the best known positions in the search-space, which are updated as better positions are found by other particles. This is expected to move the swarm toward the best solutions. The fitness is calculated as mean square errors between the trained global best weight and the given target obtained by the average weight of the field measurement of the succeed handover transactions.

PSO is originally attributed to Kennedy, Eberhart and Shi [19,20] and was first intended for simulating social behavior [21], as a stylized representation of the movement of organisms in a bird flock or fish school. We use the method to first choose which nonlinear neuron to use, and then decide what the state that represents each weight it should be.

The contributions of our work are 1) we are the first to find an important class of the neuron network that combines the memory side with the computation side, and that can mimic the cluster of composite neuron structure like our middle brain; 2） the application of the method to media independent handover for an open-flow control application;

The paper is organized as follows. Section II details for an intercerebral neural network for mobile handover controller and Section III mathematical models for the intercerebral neural network and then shows the simulation results. The last Section is the conclusion and future work.

Modeling the Nonlinear Handover Functions in Intercerebral Neural Network

The underlying nonlinearities from the mobile location to the link parameters and from the link parameters to the final handover decision are complicated. There are interactions among the link parameters of radio signal strength, packet loss, round trip time, available bandwidth, data cost and vehicle speed. Some examples are

- When the radio signal goes down, available bandwidth will go down;
- When the available bandwidth goes down, the packet loss normally goes up then;
- When the packet loss goes up, typically the round trip time will go high as well;
- When the round trip time goes up, the data cost per unit of information may rise up too.

On the other hand, there are cases

- To keep up with the vehicle speed, the data cost follows up as well;
- When the vehicle velocity speeds up, the signal strength goes down.

In another word, the link parameters are all related to each other through some complicated nonlinear relationships shown later. The only easy way to simplify these relationships is to use the intercelebral neural network that is meant to capture the multiple nonlinear relationships in order for us to make a quick handover decision from the measured field data. By using this method, we are able to capture both WiFi and GPRS characteristics at the same time. The non-linear relationships are first summarized below before we present our PSO (Particle Swarm Optimization) algorithm.

3. A MATHEMATICAL MODEL OF RADIO SIGNAL STRENGTH

The principle of RSSI (Received Signal Strength Indication) ranging describes the relationship between transmitted power and received power of wireless signals with respect to the distance among the nodes. This can be expressed in eqn. (1)

$$\log(Pr)=\log(Pt)-n^{*}\log(d) \tag{1}$$

Where, Pr is the received power; Pt is the transmitted power; d is the distance between the mobile scanner on the vehicle to the base station or the access point; and n is the transmission path factor depending on the propagation environment. Note that the typical value n for WiFi is around 3 to 4 and for GPRS is around 2 to 3.

As we can see here, just using one kind of neuron to memorize both values in a media independent mode could be difficult.

4. MATHEMATICAL MODEL FOR PACKET LOSS

We cannot obtain a closed formula to describe the packet loss because the queueing system is nonlinear. We can either use numerical calculation or approximation to calculate the packet loss. One approximation is

$$\log(Pl)=\log(Po)+m^{*}\log(\rho) \tag{2}$$

Where, Pl is the current packet loss; Po is the initial packet loss, when the system has no buffer to queue the packet; ρ is the current system load condition; and m is the equivalent buffer size, i.e. the number of average packets that the buffer can hold. Considering the fractal bursty nature of the data traffic, in practical calculation, it could be the number of the biggest packet that the mobile device can handle within the given time limit from the packet source all the way to the final destination. It is again hard to just use homogeneous neurons to capture the buffer sizing situation in an open flow fashion, typical m for WiFi is bigger than that of GPRS.

5. MATHEMATICAL MODEL OF ROUND TRIP TIME

The principle of Round Trip Time (RTT) ranging [22] describes the relationship between round trip time of wireless signals with respect to the effective distance variation among the nodes. It is shown in formula of (3):

$$RTT(n+1)=RTT(n)+\Delta d/c \quad (3)$$

Where, RTT(n+1) is the RTT for the current packet; RTT(n) is the RTT for the previous packet; and c is the speed of light. The parameter Δd is the effective distance variation between the mobile on the vehicle and the base station (or the access point), and it follows the Extended Gaussian function, shown below:

$$f(x(t))=\exp(-x^{g}/(ft))/(\pi ft)^{1/2} \quad (4)$$

Where, f(x(t)) is the probability density function for the Δd. The parameter f is the Fractal dimension of the process, when g=2, the process becomes the Gaussian process. The longer the packet travels, the closer it will be to Gaussian process. A typical value for WiFi is less than 2 and for GPRS it is more than 2. Here again just using one kind of neuron to memorize both open flow values could be difficulty.

6. MATHEMATICAL MODEL FOR AVAILABLE BANDWIDTH

The available bandwidth is given by the following formula:

$$B=C-O \quad (5)$$

Where, B is available bandwidth; O is the overhead consumed by the control and management signals or any non-white noise such as the interference; and C is the total channel capacity limit (measured in bits/second) given by the famous Shannon theory under the strictly white noise assumption. That is

$$C=W*\log(1+S/N) \tag{6}$$

Where, W is the bandwidth in Hz; S is the signal strength in watts across the bandwidth W and N is the noise power in watts across the bandwidth W.

7. MATHEMATICAL MODEL OF VEHICLE SPEED

The model for vehicle speed is simple, and given by

$$fd=V/\lambda \tag{7}$$

Where, fd is the Doppler frequency spread; V is the vehicular speed; and λ is the wireless wave length. In practice, the mobility model is divided into three categories: a) School or Residential area where the normal speed is 20 Miles/Hour; b) Urban or Business area where the typical speed is 50 Miles/Hour; and c) Highway or Rural area where the speed is 80 Miles/Hour. Take highway as an example, the value for our dual mode mobile scanner is 106Hz at 900MHz GPRS, and 212Hz at 1800MHz. At 2.4GHz WiFi, it is 283Hz. One can see that using only one type of neuron to memorize all the open flow values at different speed for different RF frequency could be difficult.

8. MATHEMATICAL MODEL FOR DATA COST

There are not so many formula for data pricing for the current GPRS network. Operators normally charge usage by the connection time in seconds. For WiFi network, operators usually charge usage by the total volume in bytes. Either way, the pricing strategy mainly follows the well know Pareto distribution:

$$Pc=(Xm/x)^{\alpha} \tag{8}$$

Pc is probability of the cost of the service, where Xm is the minimum possible value of x, representing either connection time or volume, and α is a parameter with a positive value. The Pareto Type I distribution is characterized by a scale parameter Xm and a shape parameter α, which is also known as the tail index or the Pareto index. The maximum profit strategy is usually done in such as way that the profits collected from each segmented user groups are relatively equal.

For future cooperative network, the operators may introduce more complicated pricing models (such as charging by the location, by the transmitting power, by the rush hour, or by the hot spot, small cel etc.) in order to maximize the profit. Since the relationship between the cost and the location is nonlinear; using the nonlinear neuron network to guide the open flow handover is always beneficial.

9. PARTICLE SWARM OPTIMIZED INN FOR DECISION WEIGHT IDENTIFICATION

In an Artificial Neural Network, simple artificial nodes, known as "neurons", "processing elements" or "nonlinear units", are connected together to form a network which mimics a biological neural network.

There is no single formal definition of what an artificial neural network is. However, a class of statistical models may commonly be called "Neural" if they possess the following characteristics:

1) Consist of sets of adaptive linear weights, i.e. numerical parameters that are tuned by a learning algorithm, such as PSO, and

2) Capable of approximating non-linear functions of their inputs, using different nonlinear functions, such as RBF or SBF.

The adaptive weights are conceptually connection strengths between neurons, which are activated during training process, such as PSO.

Particle Swarm Optimization (PSO) is a computational method that optimizes a problem by iteratively trying to improve a candidate solution with regard to a given measure of quality. PSO optimizes a problem by having a population of candidate solutions, here dubbed particles, and moving these particles around in the search-space according to simple mathematical formulae over the particle's position and velocity. Each particle's movement is influenced by its local best known position but, is also guided toward the best known positions in the search-space, which are updated as better positions are found by other particles. This is expected to move the swarm toward the best solutions. PSO is originally attributed to Kennedy, Eberhart and Shi [19].

The PSO trained INN decision model is shown as Fig. 2. Where Logsig () is the Sigmoid Based Function (SBF) neuron, and Exp(Abs()) is our Radial Based Function (RBF) neuron, yt is the vector contains 6 weights listed in Table II. During the course of computer simulation the mobile parameters are as follows: QVGA display, 2 Mbit pixel camera, one bar code per second, 512 Mb memory; the average driving speed is 50 km/hr, one trip per day, from Monday to Friday, total five trips. The sample data for learning is measured from the service truck made by the M2M technology Ltd (a subsidiary of GenieView Inc) for Jufeng Group. All the data parameters are normalized for ease of Matlab simulations, i.e. the trained input and output sample data is normalized to accelerate the convergence of neural network optimization process. This neural network consists of one hidden layer, including 6 neurons.

All of the hidden layer neurons adopt either traditional sigmoid or advanced radial based function. This neural network has only one output vector scale, contains the 6 decision weights. Input and output layer neurons adopt pure linear transformation function. For PSO training, the error between the trained neural network output and the expected value from the field testing with the error variation acts as the fitness function. Particle size is 20, mutation probability is 0.2, and both initial position and velocity are of zero. The sum of error square is 0.25% when the neural network is trained less than 64 steps. It is shown that the optimization process is convergent in the course of computer simulation. Thus a PSO trained neural network module is produced to realize the nonlinear mapping relationship between the measured parameters of mobile in the driving vehicle system.

The input channel of the trained neural network module is fed with the current link parameters when the system is in motion as shown in Figure 2. The neural network output is produced by the weights trained with computer simulation by the measured field data.

During the training period, PSO algorithm searches for the best solution by making particles moving around the search space according to the global best particle. But when one particle is selected as the global best particle continuously, the other particles may converge at the global best particle repeatedly, which makes the particle swarm fall into local optimization. Especially, for the complex interacting of multiple parameters here, PSO algorithm tends to be premature, so it may not guarantee the convergence to the most advantaged set of INN coefficients.

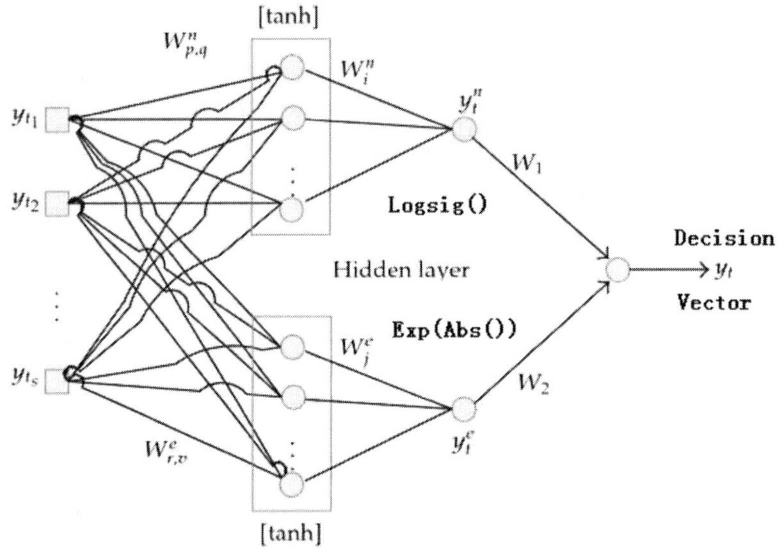

Figure 2: Neural network maps link parameters to handover decision.

To overcome the above-mentioned pitfall, the combined algorithm called AMPSO (Particle Swarm Optimization with Adaptive Mutation) is employed as a part of the Genetic Algorithm. Led by nature and the basic principles of physics in a changing environment, AMPSO has the capability to increase population diversity with respect to the genetic variation. It overcomes the problem of premature convergence phenomenon [11], which means the search ends at local minimum, mutation will restarts the search process. To gain more confidence on the INN, we have also trained it by the traditional BP (Backward Propagation). We found that by carefully changing the initial conditions, we can end up with the same tables I as shown below

Here is the pseudo code of the algorithm:

1) Initialization
2) Call the fitness calculation subroutine
3) If the error goal is met?
4) No: update the particle position and speed
5) If premature?
6) Yes: mutate the particle, Go back to to 2)
7) No: go back to step 2).
8) Yes: Training ends.

Figures 3-5 show the results of computer simulation for one week of working period. The neural network output tracks the measured output of the controller model closely. The training error is set to less than 0.25%. The simulation is carried out with five days in a cycle; the individual days are plotted on the same graph for easier visual comparison, between different combination of log sigmoid and absolute exponential radial activations.

The upper sub-graph of figure 3-5 represents the fitness value or training errors, the smaller the better accuracy. The lower sub-graph of them is the actual weight for each link parameters of each day, the more consistent across each day, the better coverage.

From which, we can see that the Figure 4 presents the best result, where the tracking error is the minimum, and the weights are pretty consistent across the week (covers 5 working days). Figure 6 presents Weight optimized via 3 SBF & 3 RB.

Tables I and II compare the performance of different combinations of RBF and SBF neurons in the hidden layer. Let RxSy to represent the case of using x RBF neurons followed by y SBF neurons in the hidden layer. One can see that the case R3S3 has the smallest weighting error of only 1.5% indicating that it can converge closer to the theoretical value than the other cases. This is very important observation for the implementation of the algorithm for a number of semi random trips where we would like to a robust neuron structure that can fit all situations. R3S3 is such good candidate because the training error is low, and the trained weight is consistent.

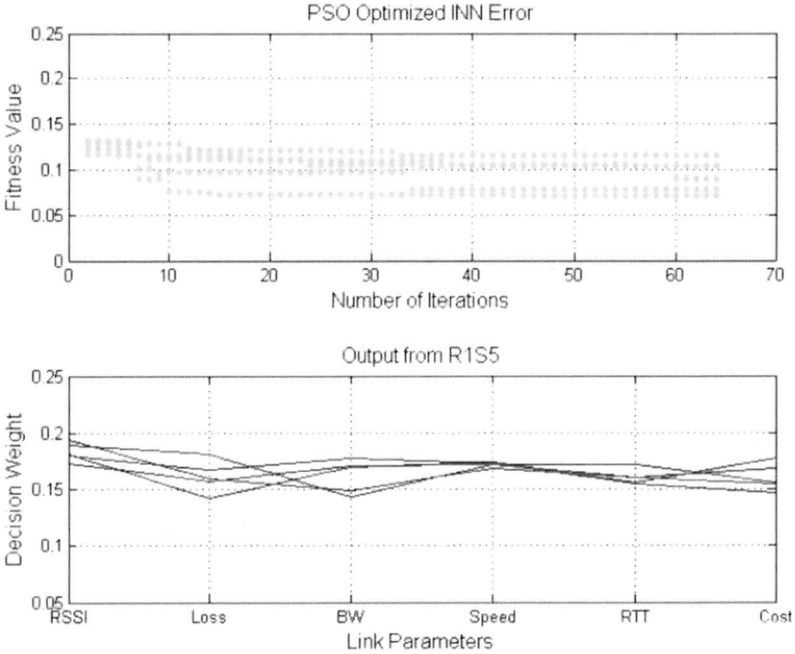

Figure 3: Weight optimized via 1 RBF and 5 SBF.

From above table, we can see the best fit case is six times better than the worst fit case in terms of the accuracy; and the best fit case is four times better than the worst case in terms of the daily standard deviation.

From above results, we see that the intercerebral method offers the additional flexibility to fit the complicated control of handover situations. Due to its extra nonlinear decision regions, it allows the search to be carried out in the WiFi domain and GPRS domain in parallel, while single function can only fit either one at a time, not both.

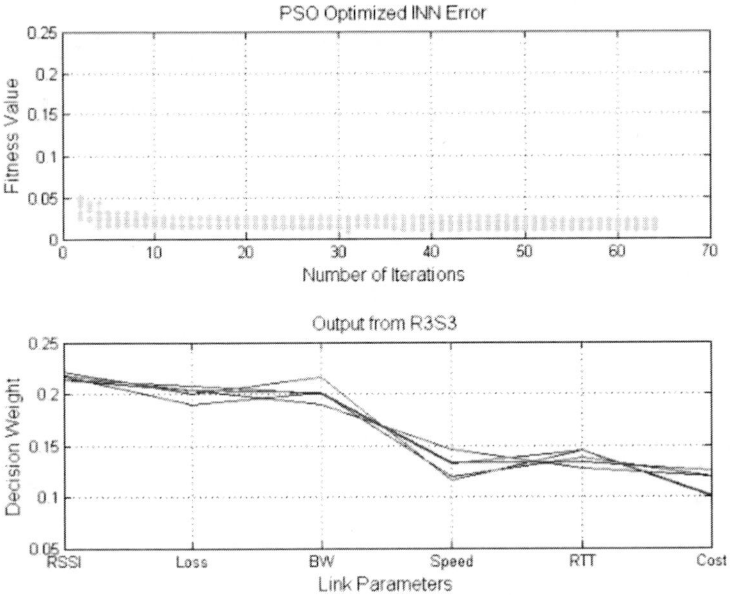

Figure 4: Weight optimized via 3 RBF and 3 SBF.

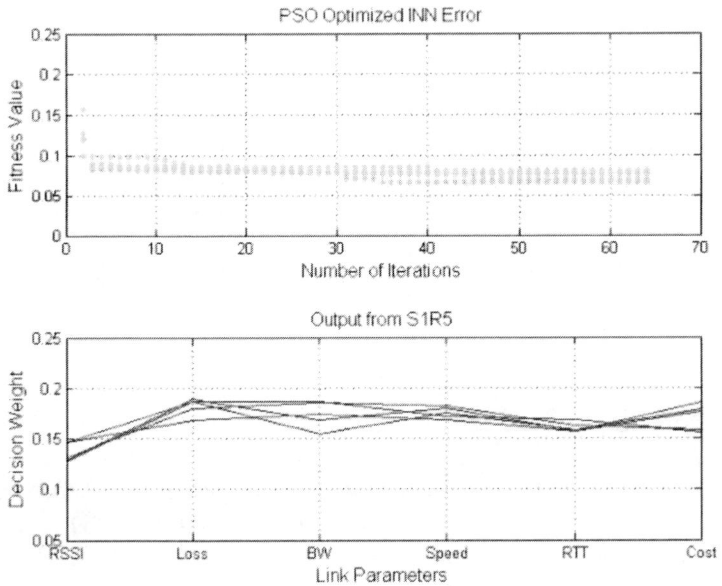

Figure 5: Weight optimized via 1 SBF and 5 RBF.

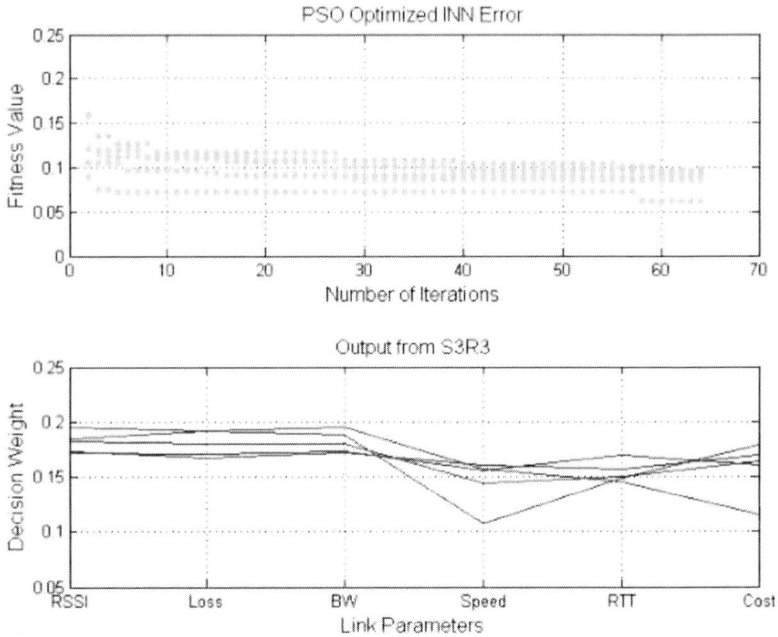

Figure 6: Weight optimized via 3 SBF and 3 RB.

Table 1: HBF Error with Different Combination.

Merits Optimized	R1S5	R3S3	S1R5	S3R3
Final Error	0.09	0.015	0.074	0.08
Daily Deviation	0.02	0.005	0.008	0.017

10. CONCLUSIONS

The eight equations of a media independent handover for fast delivery logistical vehicle information system that is controlled by open flow controller, is obtained in geographical coordinates system, that takes vehicle speed, route etc into considerations. An intercerebral neural network trained by particle swarm optimization with genetically mutated algorithm identifies the link going down indicate from radio signal strength, packet loss, round trip time, available bandwidth, data cost, vehicle speed based on field test data. The simulation shows that the proposed approach is a viable engineering solution towards the low cost high volume and precise controlling of the media independent handover system. New algorithm makes the trained decision more consistent with each other; in other words, it minimizes the manufacture cost of each smart scanner goes with dispatching vehicle. The only disadvantage is that the initial amount of statistics needed from the field test could be huge, to reach certain level of accuracy; in the moment, the statistics came from the limited routes road test, stored on the cloud servers during service trail period (www.vip56.cn). This paper is to establish a proof of concept and baseline for handover control, by intercerebral neuron methodology. Our simulation indicates that the amount of user satisfaction in terms of daily deviation can be improved is around 37%.

Table 2: HBF Application with Different Combination

Structure Weights	RSSI	Loss	BW	Speed	RTT	Cost
R1S5	0.18	0.16	0.17	0.17	0.16	0.16
R3S3	0.22	0.2	0.19	0.14	0.13	0.12
S1R5	0.15	0.18	0.17	0.17	0.16	0.17
S3R3	0.19	0.18	0.18	0.14	0.16	0.15
Field	0.21	0.19	0.18	0.15	0.14	0.13

REFERENCES

1. Dai S, Zhuang P, Xiang W(2013) GSO: An Improved PSO Based on Geese Flight Theory. Advances in Swarm Intelligence, Harbin, China.

2. Balasubramanian N, Balasubramanian A, Venkataramani A (2009) Energy consumption in mobile phones: a measurement study and

implications for network applications. Proceedings of the 9th ACM SIGCOMM conference on Internet measurement. ACM, New York, USA.

3. Li C S, Liao W (2013) Software defined networks. J Comm Mag, IEEE, 51: 113-113.

4. Goth G (2011) Software-Defined Networking Could Shake Up More than Packets. J Internet Comput, IEEE, 15: 6-9.

5. Taniuchi K, Ohba Y, Fajardo V (2009) IEEE 802.21: Media independent handover: Features, applicability, and realization; J Comm Mag, IEEE, 47: 112-120.

6. Taylor K, Jadeja JC (2005) Computer aided design and manufacturing: analysis and development of research issues. International Society for Optics and Photonics: 59990Z-12.

7. Cus F, Zuperl U, Kiker E (2006) Adaptive controller design for feedrate maximization of machining process. J Achiev Mat ManEng, 17: 1-2.

8. Yousaf M, Bhatti S, Rehan M (2009) An intelligent prediction model for generating LGD trigger of IEEE 802.21 MIH. Magazine of Emerging Intelligent Computing Technology and Applications. Springer Berlin Heidelberg

9. Oh S, Hori Y (2007) Parameter Optimization for NC Machine Tool Based on Golden Section Search Driven PSO. IEEE International Symposium on Industrial Electronics, Tokyo.

10. Das Subir, Oba Yoshihiro. MIH pre-authentication: U.S. Patent 8,036,176. 2011-10-11.

11. Jana ND, Tapas Si, Jaya Sil (2012) Particle Swarm Optimization with Adaptive Mutation in Local Bestof Particles. 2012 International Congress on Informatics, Environment, Energy and Applications-IEEA, USA.

12. Jinkun L (2013) Radial Basis Function (RBF) Neural Network Control for Mechanical Systems. Springer.

13. Neruda R, Kudova P (2005) Learning methods for radial basis function networks. Fut Gen Com Learn Sys App, 21: 1131-1142.

14. Broomhead DS, Lowe D (1988) Radial basis functions, multi-variable functional interpolation and adaptive networks. Royal Signals And Radar Establishment Malvern (United Kingdom), 3: 321–355.

15. Ho D, Boyes K, Donohoo S (2004) Numerical flow analysis for spillways. Ancold Bulletin, 4: 127-138.

16. Chatila J, Tabbara M (2004) Computational modeling of flow over an ogee spillway. Computers & structures, 82: 1805-1812.

17. Anticevic A, Cole MW, Repovs G, Savic A, Naomi R et al. (2013) Connectivity, pharmacology, and computation: toward amechanistic understanding of neural system dysfunctionin schizophrenia, Frontiers in Psychiatry.

18. Biswal BB, Mennes M, Zuo XN (2010) Toward Discovery Science of Human Brain Function. Proceedings of the National Academy of Sciences, 107: 4734-4739.

19. Shi Y, Eberhart RC (1998) Parameter selection in particle swarm optimization. Evolutionary Programming VII. Springer Berlin Heidelberg, New York, USA.

20. Bratton D, Kennedy J (2007) Defining a standard for particle swarm optimization. Swarm Intelligence Symposium, SIS IEEE: 120-127, Honolulu, HI.

21. Pedersen MEH, Chipperfield AJ (2010) Simplifying particle swarm optimization. App Soft Comp, 10: 618-628.

22. Renker G, Fairhurst G (2011) Sender RTT Estimate Option for the Datagram Congestion Control Protocol (DCCP)IETF RFC 6323.

Chapter 2

An application of swarm intelligence for the design of IIR digital filters

Ranjit Singh Chauhan1, * and Sandeep Kumar Arya2

1 Department of Electronics and Communication Engineering, JMIT Radaur, Dist. Yamuna Nagar, Haryana 135133, India

2 Department of Electronics and Communication Engineering, Guru Jambheshwar University of Science and Technology, Hisar, Haryana 125001, India

ABSTRACT

This paper presents the swarm intelligence (SI)-based particle swarm optimisation (PSO) in context of designing infinite-impulse response (IIR) digital filters. IIR filters are the part of digital filters with recursive responses. Since the error surface of IIR digital filters is generally non-linear and multimodal, global optimisation techniques are required in order to avoid local minima. The particle swarm optimisation algorithm is then applied for calculating optimal coefficients of IIR digital filters and comparison is done preferably with filter design tool and other heuristic techniques. The simulation results of benchmark filter are discussed in this paper and can be efficiently used for IIR digital filter design.

KEYWORDS

digital signal processing; DSP; digital filter; infinite-impulse response; IIR; genetic algorithm; GA; simulated annealing; SA; swarm intelligence; SI; particle swarm optimisation; PSO.

1. INTRODUCTION

Swarm intelligence (SI) is a new technique of artificial intelligence (AI) which is inspired by real world insects. There comes a lot of swarms under this category the social insects such as ants, termites, bees, and wasps and by swarming, flocking, herding, and shoaling phenomena in vertebrates. SI is recently invented high performance technique for optimisation which is objective of all the engineering application. Beni and Wang introduced the expression 'SI' in 1989, in the context of cellular robotic systems. SI, one of the most recent sub fields of artificial life, can be defined as any attempt to design algorithms or distributed problem-solving devices inspired by the collective behaviour of social insect colonies and other animal societies (Tarasewich and McMullen, 2002). SI is an innovative computational and behavioural metaphor for solving distributed problems that originally took its inspiration from the biological. While nature remains a fundamental source of inspiration for researchers in SI, new ideas originating from the most different areas in engineering and computer science are emerging and strongly influencing the field. Despite this continuous evolving of the SI definition, key principles such as self-organisation, distributions, parallelism, and exploitation of local communication mechanisms among relatively simple individuals are emerging as invariants of this innovative computational and behavioural metaphor. Examples of systems like this can be found abundant in nature, including ant colonies, bird flocking, animal herding, honey bees, bacteria, and many more. SI models have many features in common with evolutionary algorithms (EAs). Like EA, SI models are population-based (Zhou and He, 2007). The system is initialised with a population of individuals (i.e., potential solutions). These individuals are then manipulated over many generations by ways of mimicking social behaviours of insects or animals, in an effort to find the optima. Unlike EA, SI models do not use evolutionary operators such as crossover and mutation. A potential solution simply 'flies' through the search space by modifying itself according to its relationship with other individuals in the population and the environment.

Digital filtering is one of the most important tools of digital signal processing (DSP). Digital filters are capable of performance specifications that would, at best, be extremely difficult, if not impossible, to achieve with the help of an analogue implementation. Furthermore, the characteristics of a digital filter can be easily changed under software control. Digital filter can be broadly classified into two groups: recursive [infinite-impulse response (IIR)] and non-recursive [finite-impulse response (FIR)] (Hong et al., 2009). An IIR filter can provide a much better performance than the FIR filters having the same number of coefficients (Proakis and Manolakis, 2007; Ingle and Proakis, 2004). However, IIR filters might have a multi-modal error surface. Therefore, a reliable design method proposed for IIR filters must be based on a global search procedure. Digital IIR filters are widely used in the fields of automatic control, telecommunications, speech processing etc. There are two main methods for IIR digital filters design. As the error surface of IIR digital filters is usually non-linear and multimodal, conventional gradient-based design methods may easily get stuck in the local minima of error surface (Antoniou, 2005). Recently EA (Goldberg, 2005; Engelbrecht, 2005; Dai et al., 2010), such as artificial bee colony (ABC), particle swarm optimisation (PSO), ant colony optimisation (ACO), genetic algorithm (GA), simulated annealing (SA), tabu search (TS), etc., have attracted a lot attention for global optimisation problems.

Analytical or simple iterative methods normally lead to sub-optimal designs. Consequently, there is a great need of optimisation methods (heuristic type) that can be used to design digital filters that would satisfy prescribed specifications (Skaf and Stephen, 2008; Vaccaro and Harrison, 1996; Zhang and Iwakura, 1996; Argenti and Del Re, 1998). SI has become a research interest to many research scientists from various areas in recent years. The SI can be defined as any attempt for designing algorithms or distributed problem-solving devices inspired by the collective behaviour of insects and other animal societies. PSO, which was originally proposed by Kennedy and Eberhart (Fang et al., 2006), is a novel algorithm totally inspired by birds flocking in the sky or fish schooling in the sea. Benvenuto et al. (1992) described the salient features of using a SA algorithm in the context of designing digital filters with linear phase digital filter. The algorithm is then applied to the design of FIR filter. The result was not impressive. Moreover, it is computationally very expensive. Liang et al. (2003) used GA to design 1-D IIR filter with canonical-signed-digit coefficients restricted to low pass (LP) filter. Ahmad and Antoniou (2006) gave FIR filters and equalisers through the use of GA. Consequently GA requires a long computation. Oliveira et al. (2007) presented an innovative approach for designing linear FIR filters by using non-linear stochastic global optimisation based on SA techniques. Jung et al. (2008) found the design method of a linear phase finite word length finite-duration impulse response (FIR) filter through SA. Apaydin (2009) created a development of reduced-delay FIR digital filters with amplitude requirements using equiripple passband, peak constrained least squared stopband, and nearly constant group delay in passband for real time applications. Yong and Chen (2010) proposed a numerical optimisation algorithm for FIR digital filter based upon the SA paradigm and its implementation is known as fuzzy adaptive SA. Chen and Luk (2010) provided a powerful approach for solving a variety of practical signal processing problems. This contribution involves system identification application. Gao and Liao (2011) explored a novel approach of tuning PID controller which is proposed by the improved PSO. Coelho et al. (2012) concluded that the term 'SI' is used to describe algorithms and distributed problem solvers inspired by the collective behaviour of insect colonies and other animal societies. The basic drawback of preceding design methods is that the large amount of computation time is required. It is clearly indicated that all the above methods can mainly be used to design FIR digital filters. The outline of this paper is to introduce SI as an adaptive learning tool for the design of IIR digital filters. The proposed algorithm has strong search capability and is superior to the SA and GA. To test the optimisation procedure, the proposed algorithm is implemented in MATAB and results are found to be very encouraging. In this paper, articles are organised as follows: In Section 2, a brief introduction to SI approach is reported. PSO related to IIR digital filter design is given in Section 3. Two designed examples are used to verify the proposed method and simulation results of experiments are illustrated in Section 4. The conclusion and future scope is discussed in end of this paper.

2 SWARM INTELLIGENCE

SI is an AI technique that focuses on studying the collective behaviour of a decentralised system made up by a population of simple agents interacting locally with each other and with the environment. Although there is typically no centralised control dictating the behaviour of the agents, local interactions among the agents often cause a global pattern to emerge. Two of the most successful SI techniques currently in existence are ACO and PSO. ACO is a metaheuristic optimisation algorithm that can be used to find approximate solutions to difficult combinatorial optimisation problems. The ant system is a general-purpose heuristic algorithm, which can be used to solve diverse combinatorial optimisation problems. This work has been lead by Marco Dorigo at the Politecnico di Milano (Tarasewich and McMullen, 2002; Gueret et al., 2012). Ant algorithm is multi agent system in which the behaviour of the each single agent is called artificial ant, is inspired by the behaviour of real ants. ACO is a paradigm for designing metaheuristic algorithms for combinatorial optimisation problems. PSO is a global minimisation technique for dealing with problems in which a best solution can be represented as a point or surface in an n-dimensional space.

PSO is an EA developed by Eberhart and Kennedy in 1995 (Luitel and Venayagamoorthy, 2008). It is a population-based search algorithm and is inspired by the observation of natural habits of bird flocking and fish schooling. PSO is a flexible, robust population-based stochastic search/optimisation technique with implicit parallelism, which can easily handle with non-differentiable objective functions, unlike traditional optimisation methods (Yao et al., 1999). It is developed through simulation of bird flocking in multidimensional space. Bird flocking optimises a certain objective function. Each particle knows its best value so far (*pbest*). This information corresponds to personal experiences of each particle. Moreover, each particle knows the best value so far in the group (*gbest*) among *pbests*. Namely, each particle tries to modify its position using the following information:

- the distance between the current position and *pbest*
- the distance between the current position and *gbest*.

Each individual is treated as a volume-less particle in the D-dimensional space, with the position and velocity of ith particle represented as $X_i = (X_{i1}, X_{i2},.........,X_{iD})$ and $V_i = (V_{i1}, V_{i2},.........,V_{iD})$. The particles move according to the following equation:

$$V_{id} = w * V_{id} + c_1 * r_1 * (P_{id} - X_{id}) + c_2 * r_2 * (P_g - X_{id})$$

$$X_{id} = X_{id} + V_{id}$$

where c_1 and c_2 are acceleration coefficients, r_1 and r_2 are two random numbers. Vector $P_i = (P_{i1}, P_{i2},.........,P_{iD})$ is the best previous position (the position giving the best fitness value) of particle *i* called *pbest*, and vector $P_g = (P_{g1}, P_{g2},.........,P_{gD})$ is the position of the best particle among all the particles in the population and called *gbest*. Parameter *w* is the inertia weight introduced to accelerate the convergence speed of the PSO.

PSO is similar to GA in that the system is initialised with a population of random solutions. It is unlike a GA, however, in that each potential solution is also assigned a random velocity, and the potential solutions, called particles, are then 'flown' through the problem space (William and Miller, 2004; Rout et al., 2012). Each particle keeps track of its coordinates in the problem space which are associated with fitness (best solution) it has achieved so far. The fitness value is also stored which is called *pbest*. Another 'best' value that is tracked by the *global* version of the particle swarm optimiser is the overall best value, and its location, obtained so far by any particle in the population. This location is called *gbest*. The process for implementing PSO is as follow:

- Initialise a population (array) of particles with random position and velocities on *d* dimensions in the problem space.

- For each particle, evaluate the desired optimisation fitness function in *d* variables.

- Compare particle's fitness evaluation with particle's *pbest*. If current value is better then *pbest*, then set *pbest* location equal to the current location in *d* – dimension space.

- Compare fitness evaluation with the population's overall previous best. If current value is better then*gbest*, then reset *gbest* to the current particle's array index and value.

- Change the value and position of the particle according to above equations.

- Loop to step 2, until a criterion is met, usually a sufficiently good fitness or a maximum number of iterations (generation).

The pseudo code for PSO is represented in Figure 1.

The advancement in integrated circuit technology results from the increasing computing power, the simulation of evolutionary systems is becoming more and more tractable. Most applications of EAs are to the solution of static problems. Many real-world systems, however, change state frequency. These system state changes result in a requirement for frequent, sometimes almost continuous, reoptimisation. It has been demonstrated that PSO can be successfully applied to tracking and optimising dynamic systems (Zhou and He, 2007; Lai and Lin, 2012).

```
procedure PSO
    Initialization;
    while condition not met
    for each particle i (i = 1, 2........M)
    velocity update (equation 1)
    position update (equation 2)
    end for
    end while
end procedure
```

Figure 1 Pseudo code for PSO

3 PSO FOR DESIGNING IIR DIGITAL FILTERS

In general, digital filters are classified as recursive and non-recursive filters. The response of recursive or IIR filters is dependent on one or more of its past output. IIR digital filters have lot of advantages over non-recursive filters and found to be more effective. It has a much sharper transition characteristic for a given filter order. Like analogue filters with poles, an IIR filter usually has non-linear phase characteristics. If such filter subjected to an impulse then its output need not necessarily become zero. The IIR of such a filter implies the ability of the filter to have an IIR. This indicates that the system is prone to feedback and instability (Tang et al., 1998; Abdesselam, 2000; Tseng and Pei, 2001).

Consider the IIR filter with the input-output relationship governed by:

$$y(k) + \sum_{i=1}^{M} b_i y(k-i) = \sum_{i=0}^{L} a_i x(k-i)$$

where $x(k)$ and $y(k)$ are the filter's input and output, respectively, M ($\geq L$) is the filter order. The transfer function of this IIR filter can be determined as:

$$H(z)=\frac{A(z)}{B(z)}=\frac{\sum_{i=0}^{L} a_i z^{-i}}{1+\sum_{i=1}^{M} b_i z^{-i}}$$

These parameters a_0, a_1, a_2,....,a_L, b_1, b_2,....,b_M appearing in equation (3) and equation (4) are the filter coefficients. These coefficients determine the characteristics of the IIR digital filters. The digital filters have various steps in their design. In the first step, specifications of the designed filter are to be mentioned. The objective of second step is to calculate the filter coefficients for transfer function $H(z)$, approximating or satisfying a given frequency response specification. In third step, transfer function will be realised to a suitable filter network or filter structure. Input data and filter coefficients are being quantised, or rounded off due to finite word length limitation of the processor on which filtering process is taking place in fourth step. If analysis does not meet, we have to re-realise (by using other structure), re-calculate (use other technique) and re-specify (changing sampling frequency). This description is then transformed in the last step. The flow chart for the design of digital filter is shown in Figure 2 (Skaf and Stephen, 2008; Cao, 2010).

An important task for the designer is to find values of a_i and b_i such that the frequency response of the filter approximates a desired characteristic while preserving the stability of the designed filter.

Hence, the design of IIR digital filter can be considered as an optimisation problem (Kacelenga et al., 1990; Benvenuto and Marchesi, 1992) of cost function $J(w)$ stated as min$J(w)$ where w = [a_0, a_1, a_2,....,a_L, b_1, b_2,...,b_M] is filter coefficient vector. The aim is to minimise the cost function $J(w)$ by adjusting w. The cost function is usually expressed as the time-averaged cost function defined by:

$$J(w)=\frac{1}{N}\sum_{k=1}^{N}\left(d(k)-y(k)\right)^2$$

where $d(k)$ and $y(k)$ are the desired and actual responses of the filter, respectively and N is the number of samples used for the calculation of cost function. The fitness value of a solution i in the population is determined by using fitness formula given as:

$$fit(i) = \frac{1}{k + J(w)i}$$

where $J(w)$ is the cost function value computed for i and k is the number of poles outside the unit circle.

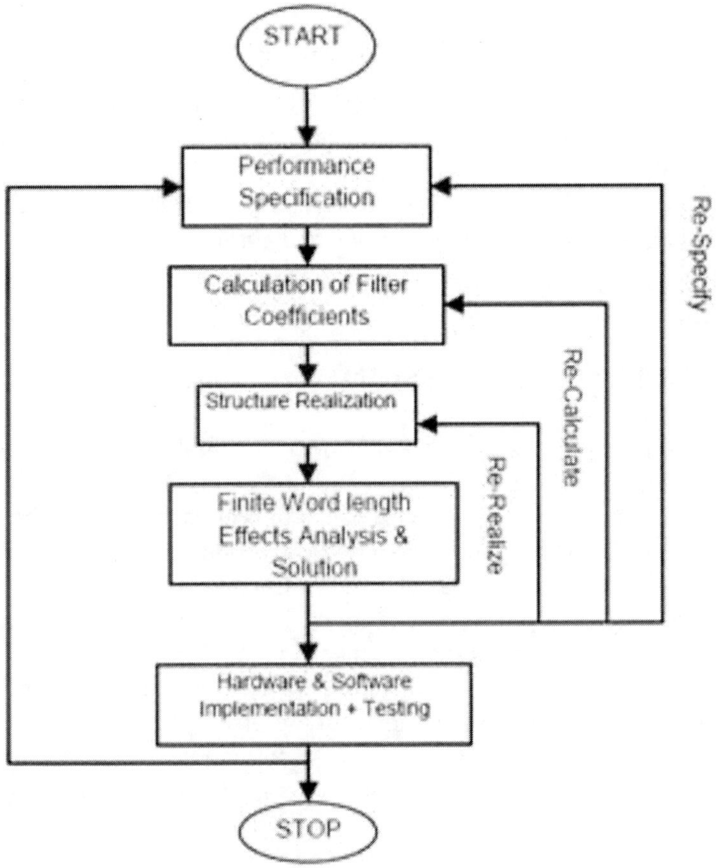

Figure 2 Flow chart of digital filter design

The error surface of digital IIR filters is generally non-linear and multimodal, so global optimisation techniques are required in order to avoid local minima. In designing IIR digital filter, the values of a_i and b_i must be such that the magnitude response of the filter approximates a desired characteristic while preserving the stability of the designed filter. Relatively little work has been published so far on PSOs applied to analogues filters. A number of practical issues are important in analogues filter design. One of them is the choice of component values (Cousseau et al., 2007; Zheng, 2003).

PSO is used to perform the design of IIR digital filter. PSO-based infinite-impulse response (PSOIIR) digital filter is implemented by MATLAB and flow chart is shown in Figure 3. The benefit of such operation is to restore the lost genetic values when the population converges too fast. The filter coefficients belonging to PSO were achieved using following parameters: with acceleration constants c_1 and c_2 each equal to 2, and the population size of 50 was assumed. As originally developed w often is decreased linearly from 1 to 0.1 in 500 iterations. The PSOIIR produces filter's coefficients that satisfy both magnitude and phase templates.

4 RESULTS AND DISCUSSION

Experiments are carried out for well known IIR digital filters (see Appendix), which have been used by many authors as a 'benchmark filter' for comparison. To compare the performance of proposed method, the results of *fda* and *SA*, GA methods (under same conditions) are also obtained through simulations (Ahn et al., 2011; Guruswamy et al., 2011). The examples were performed on Pentium IV, 2.80 GHz CPU and 1 GB of RAM. Two performance measures, i.e., mean square error (MSE) and mean standard deviation (MSD) are explored to compare the performance of proposed method, GA, SA and mean square error (*fda*)-based approaches. Simulation study is carried out in MATLAB to demonstrate the potentiality of PSO for the design of IIR digital filters. The magnitude and phase response of first example is shown in Figure 4 and Figure 5.

In Figure 6, we have depicted pole-zero behaviour of LP filter. It can be seen that the poles-zeros location of designed filter falls with in unit circle. This shows that the designed filter is also stable using PSOIIR.

The MSE and MSD are indicated in Table 1. It shows that the performance of PSOIIR gives optimal values of MSE and MSD. Figures 7 and 8 provide the qualitative measure of the performance using four algorithms.

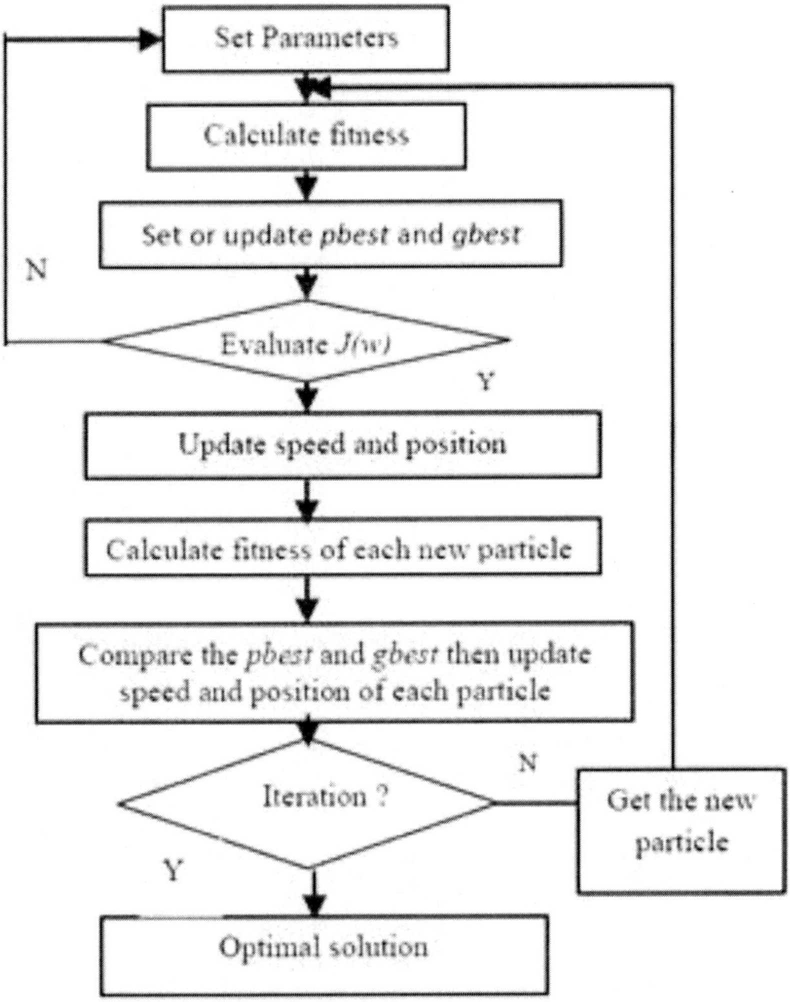

Figure 3 Flow chart of PSOIIR design

Figure 9 and Figure 10 illustrate the magnitude and phase response of second example.

Figure 11 summarised pole-zero position of high pass (HP) filter using PSO. It is clearly seen that the poles are inside the circle and zeros placement exist inside or on the verge of unit circle gives us stable HP filter.

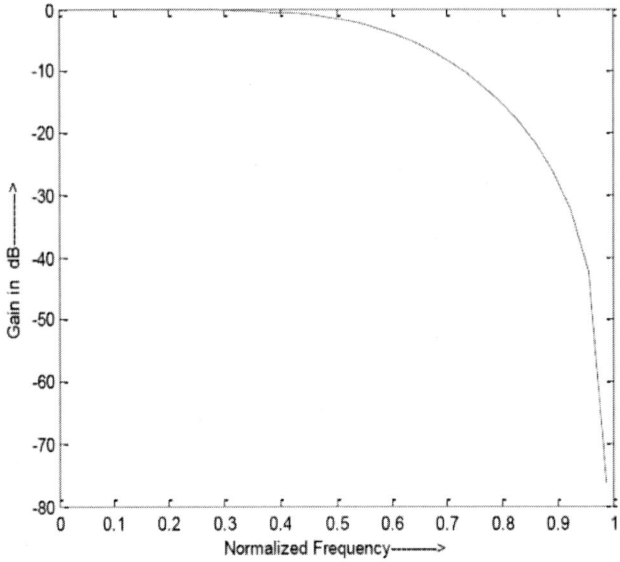

Figure 4 Magnitude response of LP filter using PSO (see online version for colours)

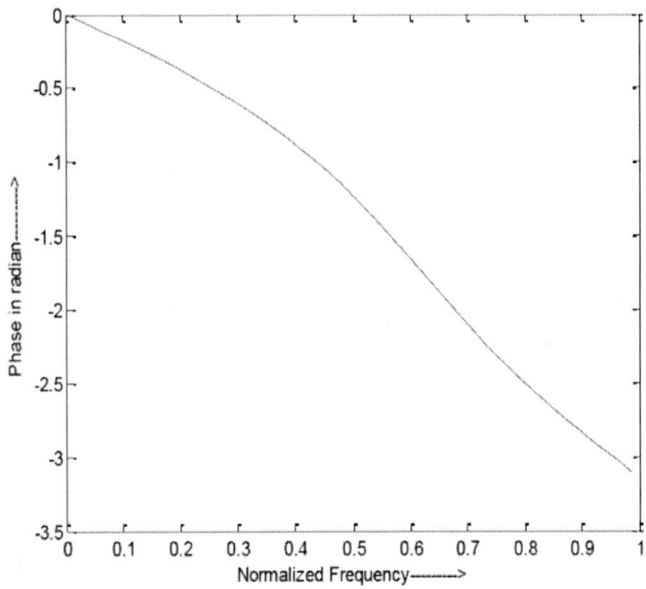

Figure 5 Phase response of LP filter using PSO (see online version for colours)

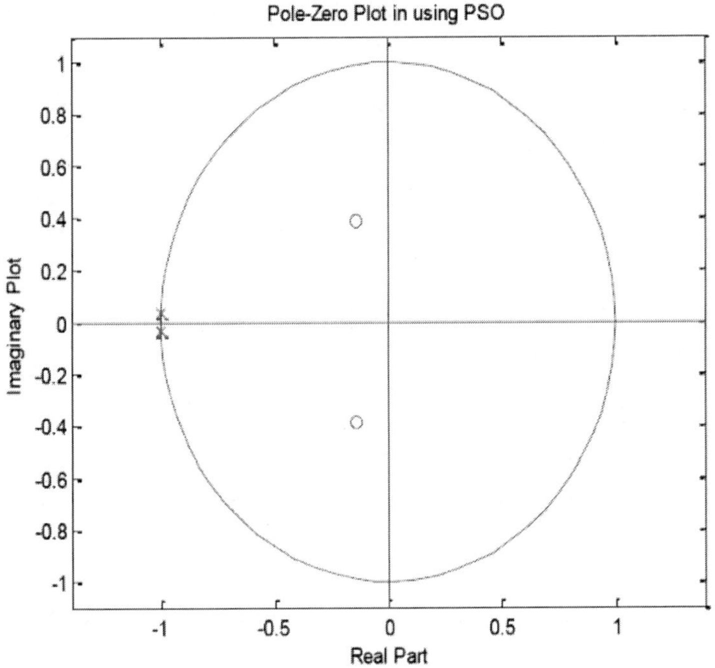

Figure 6 Pole-zero response of LP filter using PSO (see online version for colours)

Table 1. Results of MSE and MSD of LP filter

Name of method	*Order of filter*	*MSE*	*MSD*
fda tool	2	0.3296	0.3140
SA-based	2	0.3284	0.3129
GA-based	2	0.3275	0.3105
PSOIIR-based	2	*0.3195*	*0.3101*

Table 3 evaluates the values of MSE and MSD for second example under same conditions. Moreover, the MSE and MSD of second filter are shown in Figure 12 and Figure 13, respectively. Table 4 gives the coefficients of HP filter under same specifications of PSOIIR, with the order of designed filter. The designed example is compared with the other methods under same conditions (fda tool, SA and GA approaches). On comparison, it is found that the PSOIIR algorithm gives optimal coefficients for HP filter and least MSE and MSD.

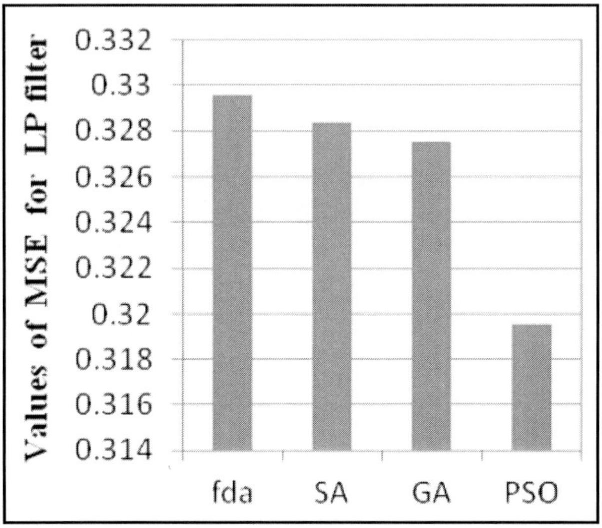

Figure 7 Comparison of MSE using various techniques (see online version for colours)

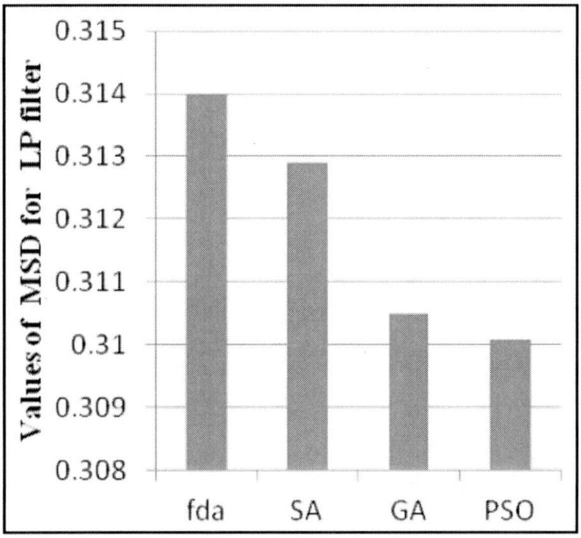

Figure 8 Comparison of MSD using various techniques (see online version for colours)

Table 2 provides a comparison of coefficients with the fda tool, SA and GA.

Table 2. Coefficients of LP filter designed with second order

Coeff.	*fda tool*	*SA-based*	*GA-based*	*PSOIIR-based*
a_0	0.3752	0.3749	0.3739	0.3659
a_1	0.7504	0.7489	0.7459	0.7313
a_2	0.3752	0.3749	0.3739	0.3659
b_1	1.0000	1.0000	1.0000	1.0000
b_2	0.3120	0.3109	0.3116	0.2916
b_3	0.1888	0.1790	0.1789	0.1723

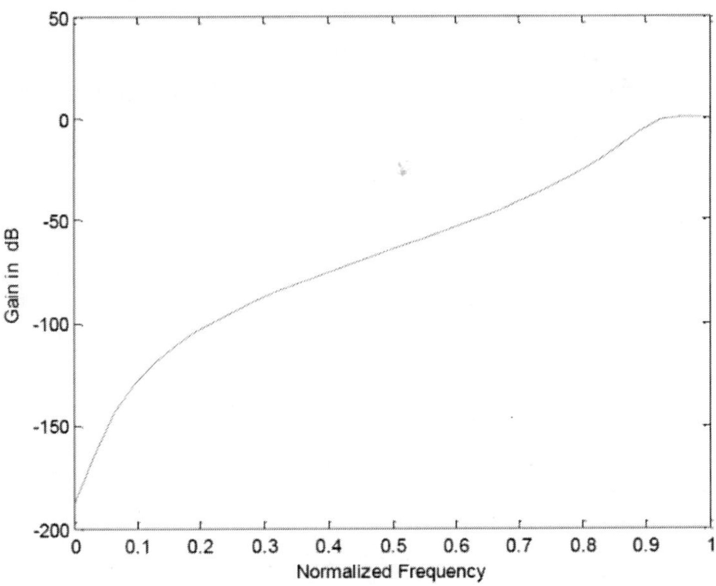

Figure 9 . Magnitude response of HP filter using PSO (see online version for colours)

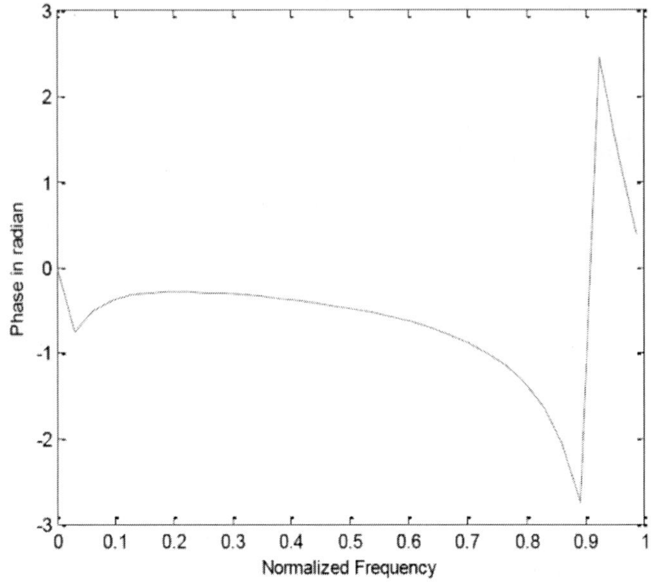

Figure 10 . Phase response of HP filter using PSO (see online version for colours)

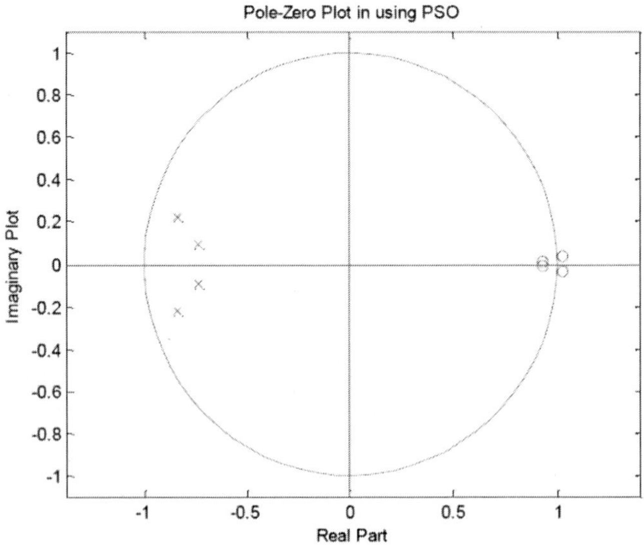

Figure 11. Pole-zero position of HP filter using PSO (see online version for colours)

Table 3. Results of MSE of HP filter

Name of method	*Order of filter*	*MSE*	*MSD*
fda tool	4	3.0589	1.4682
SA-based	4	2.9967	1.4534
GA-based	4	2.9913	1.4529
PSOIIR-based	4	*2.9817*	*1.4503*

5 CONCLUSIONS

In this paper, we developed a SI method for the design of IIR digital filters. It is concluded that PSO is a global optimisation technique for IIR digital filters, and the benefits of SI for designing digital filter have been studied. The simulation results show that PSOIIR has better, or at least equivalent, global search ability and convergence speed than others. The examples demonstrate the versatility of the proposed approach. The performance of proposed method has been compared with fda tool, SA and GA technique. Better MSE and MSD are calculated from the proposed method. The position of pole-zero of the designed filters using optimal coefficients provides us stable filters. Thus, it is believed that the PSOIIR algorithm is capable of quick and high performance. The proposed method can be extended to arbitrary magnitude response specifications and multi-band. The memetic algorithm or the other EAs can be the thrust area for the design of IIR digital filters.

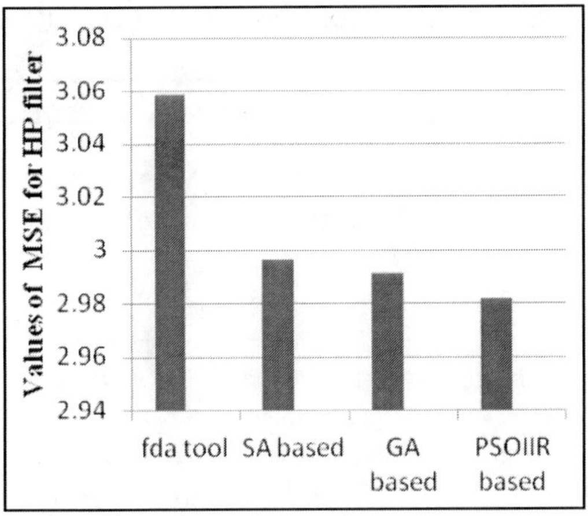

Figure 12 Comparison of MSE for HP using various techniques (see online version for colours)

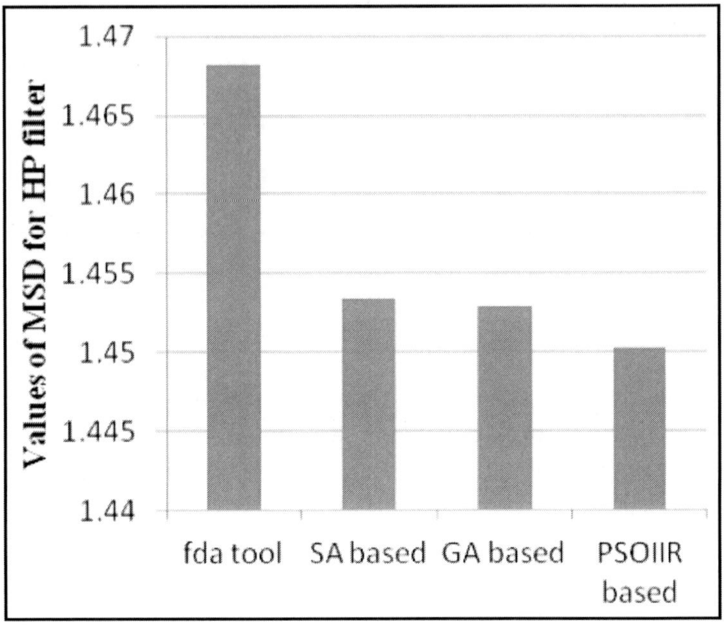

Figure 13 Comparison of MSD for HP filter using various techniques (see online version for colours)

Table 4 Coefficients of HP filter designed with fourth order

Coefficients	*fda tool*	*SA-based*	*GA-based*	*PSOIIR-based*
a_0	0.0004218886	0.0004218886	0.0004218886	0.0004218886
a_1	−0.0016875540	−0.001610240	−0.0016152230	−0.0016468845
a_2	0.0025313316	0.0022993300	0.0023143520	0.0024093298
a_3	−0.0016875540	−0.001455500	−0.0014705890	−0.0015655560
a_4	0.0004218886	0.000344561	0.0003495715	0.00038122653
b_1	1.0000000000	1.000000000	1.0000000000	1.0000000000
b_2	3.177811000	3.15992000	3.1654356000	3.1604679000
b_3	3.854654400	3.80985000	3.8067268000	3.7993156200
b_4	2.107306220	2.06993900	2.0556433000	2.0534704500
b_5	0.437002640	0.42670300	0.4196924000	0.4204098560

REFERENCES

1. Abdesselam, K.D. (2000). 'Design of stable, causal, perfect reconstruction, IIR uniform DFT filters'. IEEE Trans. on Signal Processing. 48, 4, 1110-1117

2. Ahmad, S.U. , Antoniou, A. (2006). 'Design of digital filters using genetic algorithms'. IEEE Trans. on Signal Processing. 1, 1, 1-9

3. Ahn, S. , Lee, H. , Jung, B. (2011). 'Medium PRF set selection for pulsed Doppler radars using simulated annealing'. Proceeding of IEEE Radar Conference. 23–27 May, Kansas, USA, 90-95

4. Antoniou, A. (2005). Digital Filters Analysis, Design and Application. New Delhi, Tata Mcgraw-Hill Edition

5. Apaydin, G. 'Realization of reduced-delay FIR filters for audio applications'. Elsevier Digital Signal Processing. 2009, 08, 20, 3, 620-629

6. Argenti, F. , Del Re, E. 'Design of IIR eigen filters in the frequency domain'. IEEE Trans. on Signal Processing. 1998, 06, 46, 6, 1694-1700

7. Benvenuto, N. , Marchesi, M. (1992). 'Applications of simulated annealing for the design of digital filters'. IEEE Trans. Signal Processing. 40, 2, 323-331

8. Benvenuto, N. , Marchesi, M. , Uncini, A. 'Applications of simulate annealing for design of digital filter'. IEEE Trans. on Signal Processing. 1992, 02, 40, 2, 323-332

9. Cao, L. 'Practical issues in implementing single-pole low-pass IIR filter'. IEEE Signal Processing Magazine. 2010, 11, 27, 6, 114-117

10. Chen, S. , Luk, B.L. (2010). 'Digital IIR filter design using particle swarm optimisation'. International Journal of Modelling, Identification and Control. 9, 4, 327-335 Abstract

11. Coelho, L.d.S. , Guerra, F.A. , Leite, J.V. 'Multiobjective exponential PSO approach applied to hysteresis parameters estimation'. IEEE Trans. on Magnetics. 2012, 02, 48, 2, 283-287

12. Cousseau, J.E. , Werner, S. , Donate, P.D. (2007). 'Factorized all-pass based IIR adaptive notch filters'. IEEE Transaction on Signal Processing. 55, 11, 5225-5236

13. Dai, C. , Chen, W. , Zhu, Y. (2010). 'Seeker optimization algorithm for digital IIR filter design'. IEEE Trans. on Evolutionary Computation. 57, 5, 1710-1718

14. Engelbrecht, A.P. (2005). Fundamentals of Computational Swarm Intelligence. New Delhi:John Wiley & Sons Ltd.

15. Fang, W. , Sun, J. , Xu, W. (2006). Design IIR Digital Filters Using Quantum-Behaved Particle Swarm Optimization. Berlin, Heidelberg: Springer-Verlag , 637-640

16. Gao, Z. , Liao, X. (2011). 'Design of fractional order PID controller by hybrid adaptive particle swarm optimization based on the average velocity'. Proceedings of the 30th Chinese Control Conference. 22–24 July, Yantai, sChina, 5440-5446

17. Goldberg, D.E. (2005). Genetic Algorithm in Search, Optimization and Machine Learning. Delhi:Pearson Education , Low Price Edition

18. Gueret, C. , Schlobach, S. , Dentler, K. , Schut, M. , Eiben, G. 'Evolutionary and swarm computing for the semantic web'. IEEE Computational Intelligence Magazine. 2012, 05, 16-31

19. Guruswamy, V.L. , Lang, J. , Lee, W.S. (2011). 'IIR filter models of haptic vibration textures'. IEEE Transactions on Instrumentation and Measurement. 60, 1, 93-104

20. Hong, X.U. , Gang, L.I. , Huang, C. (2009). 'A novel approach to stable IIR digital filter design'. Proceeding of IEEE Conference on ICICS. 7–10 September, Macau, China, 1-5

21. Ingle, V.K. , Proakis, J.G. (2004). Digital Signal Processing using MATLAB. New Delhi:Thomson Books

22. Jung, B.W. , Yang, H.J. , Chun, J. (2008). 'Finite word length digital filter design using simulated annealing'. Proceeding of IEEE Conference on Signals, Systems and Computers. 26–28 October, California, USA, 546-550

23. Kacelenga, R.V. , Graumann, P.V. , Turner, L.E. (1990). 'Design of filters using simulated annealing'. IEEE Proc. Int. Symp. on Circuits and Systems. New Orleans, LA, 642-645

24. Lai, X. , Lin, Z. 'Minimax phase error design of IIR digital filters with prescribed magnitude and phase responses'. IEEE Transactions on Signal Processing. 2012, 02, 60, 2, 980-987

25. Liang, L. , Ahmadi, M. , Ahmed, M. , Wallus, K. (2003). 'Design of canonical signed digital filters using genetic algorithms'. IEEE Trans. on Signal Processing. 3, 1, 2043-2047

26. Luitel, B. , Venayagamoorthy, G.K. (2008). 'Differential evolution particle swarm optimization for digital filter design'. IEEE Congress on Evolutionary Computation (CEC 2008), 3952-3960

27. Oliveira, H.A. , Petraglia, A. , Petraglia, M.R. (2007). 'Frequency domain FIR filter design using fuzzy adaptive simulated annealing'. IEEE International Symposium on Signal Processing & Information Technology, 884-888

28. Proakis, J.G. , Manolakis, D.G. (2007). Digital Signal Processing: Principles, Algorithms and Applications. 4th ed., New Delhi:Pearson Education , Low Price Edition

29. Rout, N.K. , Das, D.P. , Panda, G. (2012). 'Particle swarm optimization based active noise, control algorithm without secondary path identification'. IEEE Trans. on Instrumentation and Measurement. 61, 2, 554-564

30. Skaf, J. , Stephen, P.B. 'Filter design with low complexity coefficients'. IEEE Trans. on Signal Processing. 2008, 07, 56, 7, 3162-3170

31. Tang, K.S. , Man, K.F. , Kwong, S. (1998). 'Design and optimization of digital filter structure using genetic algorithm'. IEEE Transactions on Industrial Electronics. 45, 3, 481-489

32. Tarasewich, P. , McMullen, P.R. (2002). 'Swarm intelligence-power in numbers'. Communication of the ACM. 45, 8, 62-67

33. Tseng, C.C. , Pei, S.C. (2001). 'Stable IIR notch filter design with optimal pole placement'. IEEE Transactions on Signal Processing. 49, 11, 2673-2681

34. Vaccaro, R.J. , Harrison, B.F. 'Optimal matrix-filter design'. IEEE Trans. on Signal Processing. 1996, 03, 44, 3, 705-710

35. William, T. , Miller, W.C. (2004). 'Genetic algorithms for the design of digital filters'. Proceedings of IEEE ICSP'O4. 9-12

36. Yao, X. , Liu, Y. , Lin, G. 'Evolutionary programming made faster'. IEEE Transaction on Evolutionary Computation. 1999, 07, 3, 2, 83-102

37. Yong, Z. , Chen, W. (2010). 'Design and simulation of FIR digital filter using fuzzy adaptive method'. Proceeding of IEEE Conference on WiCOM. 23–25 September, Chengdu, China, 1-3

38. Zhang, X. , Iwakura, H. 'Design of IIR digital filters based on eigen value problem'. IEEE Trans. on Signal Processing. 1996, 06, 44, 6, 1325-1319

39. Zheng, W.X. 'Adaptive filter design subject to output envelop constraints and bounded input noise'. IEEE Transaction on Circuit & Systems-II Analog & Digital Signal Processing. 2003, 12, 50, 12, 1023-1027

40. Zhou, Y. , He, J. 'A runtime analysis of evolutionary algorithms for constrained optimization problems'. IEEE Trans. on Evolutionary Computation. 2007, 10, 11, 5, 608-620

Chapter 3

Information flow principles for plasticity in foraging robot swarms

Lenka Pitonakova[1], Richard Crowder[1], Seth Bullock[2]

[1] Institute for Complex Systems Simulation and Department of Electronics and Computer Science, University of Southampton, Southampton, UK

[2] School of Computer Science, University of Bristol, Bristol, UK

ABSTRACT

An important characteristic of a robot swarm that must operate in the real world is the ability to cope with changeable environments by exhibiting behavioural plasticity at the collective level. For example, a swarm of foraging robots should be able to repeatedly reorganise in order to exploit resource deposits that appear intermittently in different locations throughout their environment. In this paper, we report on simulation experiments with homogeneous foraging robot teams and show that analysing swarm behaviour in terms of information flow can help us to identify whether a particular behavioural strategy is likely to exhibit useful swarm plasticity in response to dynamic environments. While it is beneficial to maximise the rate at which robots share information when they make collective decisions in a static environment, plastic swarm behaviour in changeable environments requires regulated information transfer in order to achieve a balance between the exploitation of existing information and exploration leading to acquisition of new information. We give examples of how information flow analysis can help designers to decide on robot control strategies with relevance to a number of applications explored in the swarm robotics literature.

KEYWORDS

Swarm robotics Foraging, Communication, Self-organisation, Plasticity

1. INTRODUCTION

Plasticity in a robot swarm is the ability of the robots to repeatedly organise and reorganise in response to changing demand characteristics. As such, it is central to the ability of a robot swarm to cope with real-world tasks when they are extended in time or space, since these typically feature inherent dynamism in the demand characteristics that the swarm must deal with. In this paper, we study the emergence of swarm plasticity in the context of swarm foraging with homogeneous simulated swarms that need to respond to changing resource deposits. We develop an information flow measure that allows us to understand the swarm's self-organisation and present robot behavioural strategies that lead to plasticity at the level of the entire swarm.

Swarm foraging is a behaviour that has received a considerable amount of attention in swarm robotics (e.g. Krieger and Billeter 2000; Liu et al. 2007; Gutiérrez et al. 2010). Robots must find and collect resources distributed throughout their environment. While designing strategies for swarm foragers is challenging, such strategies allow swarms of robots to coordinate their exploitation of the environment and to potentially forage more effectively. Furthermore, foraging is often used as a paradigm for studying other swarm behaviours, such as task allocation (e.g. Jevtic et al. 2012) or labour division (e.g. Zahadat et al. 2013), where worksites need to be searched for in the environment and work is distributed among the robots, or agent dispersion (e.g. Ranjbar-Sahraei et al. 2012), where robots need to coordinate their movement via communication. We choose foraging as a test bed application in this paper so that our results are relevant to a spectrum of topics covered in the current swarm robotics literature.

If we could create swarms that are not only able to perform given tasks, but able to perform them autonomously and reliably over a prolonged period of time in changeable environments, robot swarms could collect garbage from our streets, extract and gather minerals or be used in logistic applications. However, it is not their ability to forage itself, but rather their ability to adapt that makes swarm foraging systems interesting and useful (e.g. Dai 2009; Ducatelle et al. 2011). Therefore, if robot swarms are to address real-life applications, we need to understand how swarm intelligence works in dynamic environments. In order to do this, we need to establish ways of analysing artificial swarms so that we can generalise findings from particular experiments and aim towards a framework or a set of design principles that can guide future research and engineering. Recent work on self-organisation in artificial swarm systems has shown an increasing demand for such generalisations (e.g. Parunak and Brueckner 2004; Serugendo et al. 2006; Winfield 2009).

In order to gain a detailed understanding of swarm foraging and identify what robot-level behaviour leads to swarm-level plasticity, we simulate and analyse foraging robot swarms of a gradually increasing complexity in a number of different environments. Our robots utilise recruitment, i.e. they inform each other about foraging locations through direct communication. Section 2 gives an overview of relevant aspects of social insect foraging behaviour and discusses state of the art in swarm robotics. We introduce our simulation environment and the robot control algorithm in Sect. 3 and evaluate performance of our robot swarms in simple static environments in Sect. 4. We then test the ability of swarms with various communication strategies to discover and discriminate between deposits of different quality in Sects. 5 and 6, and we demonstrate the importance of a balance between exploration and exploitation in dynamic environments, where deposit quality changes over time. By measuring *information value*, which captures the direct effect of information flow on resource collection, we show in Sects. 7–9 that if information spreads slowly through swarms, they are likely to remain effective as foraging conditions change, due to their ability to balance information utilisation and acquisition. On the other hand, if information spreads quickly through swarms, while performance may be improved under specific environmental conditions, such swarms may also be less capable of plastic self-organisation. The information value measure can thus help designers to decide on robot behavioural parameters that lead to desired work modes for their swarms. Furthermore, in Sect. 9, we offer a set of swarm design principles focused on communication and discuss their relevance to a wider range of problems including swarm foraging from multiple sites in parallel, swarm self-regulation, emergent task allocation and evolutionary robotics.

2. BACKGROUND

2.1 Swarm foraging in nature

The behaviour of swarms of foraging robots is often inspired by ants or bees. Colonies of these social insects demonstrate an incredible ability to self-organise when their environment changes and have therefore been studied by both biologists and engineers.

Ants use pheromone trails to indicate paths through the environment and to inform their nest mates about foraging locations (Sumpter and Beekman 2003; Arab et al. 2012). Pheromone paths that lead to better food sources and are used by more ants become stronger over time and attract more and more workers, while paths to inferior sources, that are not being reinforced frequently enough, gradually evaporate.

Bees are another example of animals that use recruitment when foraging. However, unlike ants, bees use direct signalling and have a designated area in the hive, called the *dance floor*, where they perform recruitment *waggle dances* for nest mates that are interested in foraging (von Frisch 1967; Seeley et al.1991; Biesmeijer and De Vries 2001). While the length and strength of a waggle dance is related to the quality of a particular flower patch, the position and orientation of a bee on the dance floor relative to the sun encodes the location of a patch relative to the hive, allowing recruits to travel to specific advertised food sources. An individual's decisions about whether to waggle dance, forage or abandon a patch are affected by olfactory and taste information from nectar samples obtained from other foragers through *trophallaxis*after one bee sends a *begging signal* to another (De Marco and Farina 2003; Farina et al. 2005). For example, when a forager discovers that other bees are processing nectar of a much better quality, it abandons its own source faster and has a higher preference for the better source for a number of days (De Marco and Farina 2001).

Compared to ants, bees are generally better at achieving plasticity. For example, ants find it difficult to establish a new shorter route to a food source if an established trail already exists (Ribeiro et al. 2009).

Furthermore, ants rely on pheromone evaporation, which takes time and makes diversion of foraging effort from a depleted to a new patch relatively slow. On the other hand, bees abandon patches based on individual decisions, making their responses faster when a patch is depleted (Sumpter and Beekman 2003). Bee colonies also have the ability to switch between patches when their relative quality changes (Seeley et al. 1991). It has been argued that trophallaxis plays an important role in dynamic environments, as it allows information about flower patch quality to spread through the whole hive within hours, while waggle dancing only affects bees that follow dances and is thus a slower communication method (Farina et al.2005). Additional evidence suggests that returning scouts use *stop signals* to directly inhibit waggle dances of recruiters that advertise alternative sites (Seeley et al. 2012). Bee colonies also achieve flexibility through opportunistic scouting when a recruited forager gets lost due to errors in signal propagation during waggle dance (Seeley 1994) and through *inspection*, i.e. occasional re-evaluation of abandoned flower patches (Granovskiy et al. 2012).

2.2 Robot swarm foraging in static environments

There is a considerable volume of research that concentrates on swarm foraging during which robots are solitary and do not communicate at all. Such swarms are often used in static environments to retrieve targets and bring them back to a designated location (e.g. Arkin 1992; Balch 1999; Ulam and Balch 2004).

Ant-inspired robots in simulated experiments can drop cues directly into the environment in order to help others reach items of interest (Drogoul and Ferber 1993). The use of pheromone and its evaporation and dispersion can also be simulated (Fujisawa et al. 2014). In the real world, robots that move on a phosphorescent floor can use LEDs to create glowing paths (Mayet et al. 2010). Alternatively, robots can deposit alcohol trails and use chemical sensors to follow them (Russell 1999; Fujisawa et al. 2014) or a centralised server can store virtual pheromone deposited by robots and use a projector to display it, allowing robots to follow pheromone trails using visual sensors (Kazama et al. 2005; Garnier et al. 2007). In order to avoid the difficulty of using pheromone-like substances that require a specific arena set-up, a virtual pheromone is often represented by designated stationary robots that communicate pheromone levels to others nearby (e.g. Hoff et al. 2013; Ducatelle et al. 2011).

Similarly, in a set-up inspired by bee trophallaxis, a portion of the swarm is designated for propagation of values passed from robot to robot, allowing a gradient to be established between the base and a resource patch (Schmickl and Crailsheim 2008; Nouyan et al. 2009). It is also possible to use the whole swarm as a medium that holds pheromone paths, allowing virtual ants to travel through the robots and establish the shortest path to a resource (Campo et al. 2010).

Some ant-inspired (e.g. Krieger and Billeter 2000), as well as bee-inspired (e.g. Alers et al. 2011; Lee and Ahn 2011) robotic systems use direct signals that allow robots to recruit others to specific deposit locations when they arrive at the base. Alternatively, robots may communicate at any point when they meet each other during foraging (e.g. Valdastri et al. 2006; Gutiérrez et al. 2010; Miletitch et al. 2013). To localise themselves and the objects of interest, robots use path integration and usually store one vector pointing towards the base and one pointing towards a found deposit.

Bee-inspired algorithms have also been applied to help robots aggregate in areas of interest (Schmickl and Hamann 2010) or to optimise a swarm's energy intake from resource that was collected in the environment and processed in the base (Thenius et al. 2008). It is often difficult to perform experiments with large swarm sizes, to repeat experiments a sufficient number of times, or to collect statistics about robot behaviour during real-world experiments, causing many researches to rely on simulated data when a thorough analysis of swarm behaviour is required (e.g. Lee and Ahn 2011; Liu et al. 2007; Campo and Dorigo 2007; Miletitch et al. 2013).

2.3 Robot swarm foraging in dynamic environments

When foraging in environments that change over time, swarms need to possess some form of self-organisation on the level of the collective or adaptation at the individual level in order to cope with changing foraging conditions. Bees use a combination of both techniques. They adapt their response thresholds to various stimuli (Seeley 1994) and learn odours of profitable flowers (Farina et al. 2005). Colonies are also capable of self-organisation that results from evolved responses of bees to individual and social information (e.g. Seeley et al. 1991; De Marco and Farina 2003).

A frequently studied swarm robot behaviour is the ability of swarms to adjust the number of foragers and resting robots based on changing resource abundance. Robots can be equipped with means to perceive their own foraging performance and the performance of others and adapt their own control parameters accordingly (e.g. Campo and Dorigo 2007; Liu et al. 2007). Alternatively, self-organisation on the swarm level, that emerges when robots change their actions based on individual and social information, can also be used to make swarms adapt (e.g. Sarker and Dahl 2011).

In other experiments, foraging robots were required to distribute their foraging effort between different types of resource proportional to type abundance. It was shown that self-organising swarms could successfully solve this task if a correct balance between information sharing among robots and information acquisition by robots is achieved (e.g. Jones and Matarić 2003; Schmickl et al. 2007).

Finally, some foraging experiments involve bucket brigading, where robots form chains of adjacent "working areas" that run from the deposit location to the robot base, and robots move resources along these chains (e.g. Shell and Mataric 2006; Nouyan et al. 2009; Pini et al. 2013). An individual robot's working area can be adapted online to give a swarm the ability to follow mobile deposits (Lein and Vaughan 2009).

Here, in an approach similar to that of Jones and Matarić (2003), Schmickl et al. (2007), Sarker and Dahl (2011), robots do not adapt their individual control parameters, but share information with one another in order to achieve self-organisation at the level of the swarm.

3. METHODS

3.1 Simulation environment

All the experiments reported here are performed in the ARGoS simulation environment (Pinciroli et al. 2012) using MarXbot robots (Bonani et al. 2010). ARGoS is a C++ environment with a realistic 3D physics engine that was specifically designed for program compatibility with MarXbots (also called foot-bots) and e-pucks. By using this type of simulation environment, we can model not only decision-making of robots, but also their physical interactions with each other and with the world. The physical aspect is important as it can lead to various interferences between robots and affect their performance, for example when multiple robots try to access a deposit at the same time or when they attempt to communicate with each other (Pitonakova et al. 2014).

The simulation takes place in continuous space and updates itself 10 times per second. The foraging arena is 50 m ×× 50 m large and contains a centrally located circular base surrounded by resource deposits (Figs. 1, 2). A similar set-up has been used previously in simulated (e.g. Balch and Arkin 1994; Campo and Dorigo 2007) and real-world (e.g. Labella et al. 2004; Gutiérrez et al. 2010) experiments.

The base has a radius of three metres, and is divided into two sections: an interior circular dance floor and an annular unloading area around it (Fig. 1). A light source is placed above the middle of the base that the robots can use as a reference for navigation towards and away from the centre of the base (as in, for example, Krieger and Billeter 2000; Ferrante and Duéñez Guzmán 2013; Pini et al. 2013).

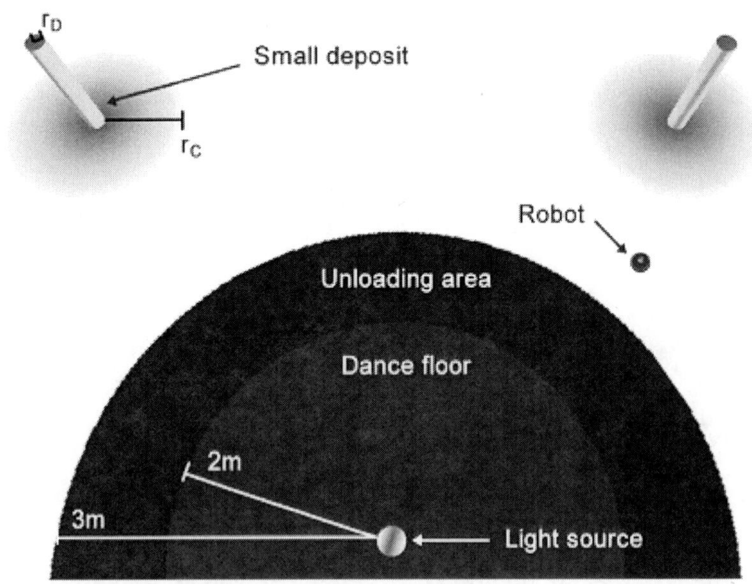

Figure 1. ARGoS simulation screenshot of a base and scattered small deposits. The base consists of a circular dance floor with a radius of 2 m, where recruitment takes place, and an unloading area that forms a ring around the dance floor where returning foragers drop collected material. There is also a light source above the centre of the base. The whole base has a radius of 3 m. A robot with radius of 8.5 cm is located near the base. Each deposit has a *colour* gradient around it to allow nearby robots to navigate towards it (Color figure online)

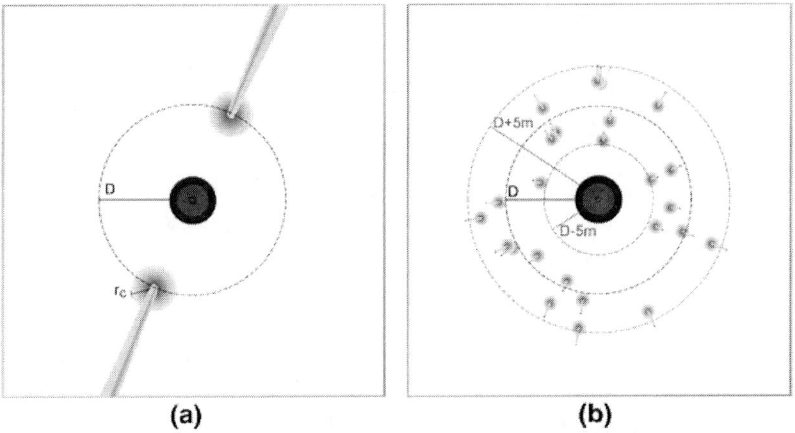

Figure 2. ARGoS simulation screenshot of the experimental arena containing a base in the centre and deposits in the **a** Heap2 and **b** Scatter25 scenarios with $D=9D=9$ m. The deposits are represented as cylinders, with their heights corresponding to their respective volumes, and have *colour* gradients with radius rCrCaround them to guide navigation of nearby robots (Color figure online)

Resource is distributed throughout the environment in the form of a number, *N*, of discrete deposits. Each deposit is cylindrical, with radius rDrD. The value of deposit *i* to a robot is related to two properties: the*volume* of the deposit, ViVi, which represents the gross amount of material it contains, and the *quality* of the deposit, QiQi, which represents how rich this material is. The total amount of resource in a deposit is thus $Vi \times QiVi \times Qi$. Each robot loads a maximum volume of LmaxLmax units per foraging trip. The total amount of resource that a robot may load from deposit, *i*, in a single trip is thus $L=Lmax \times QiL=Lmax \times Qi$ or $L=Vi \times QiL=Vi \times Qi$ if $Vi<LmaxVi<Lmax$.

Consider two deposits, *A* and *B*, with volumes greater than LmaxLmax and where deposit *A* has twice the volume of *B*, but deposit *B* has twice the quality of deposit *A*. Robots carrying all of deposit *A* back to the base will have foraged the same amount of resource as robots carrying all of deposit *B* back to the base because $VA \times QA=VB \times QBVA \times QA=VB \times QB$, but robots foraging from deposit *A* will have needed more trips to achieve this since $VA>VBVA>VB$.

In the following environments, the volume of each deposit is 100/*N* units, i.e. total volume of deposits is 100 units and is conserved across environments. *N* is always selected so that ViVi is an integer. The default maximum value of *L* is Lmax=1Lmax=1 unit, although LmaxLmax is reduced to 0.25 units in later simulations in order to increase the amount of time taken by swarms to complete the foraging task. The default value of *Q* for all deposits is 1, although some scenarios are specifically designed to contain deposits of varying quality.

In order to enable robots close to a deposit to move towards it, a colour gradient with radius rCrC is centred on the floor around each deposit. In a real-world experiment, navigation based on the colour gradient could be replaced by visual-based navigation, for instance.

There are two types of scenario (Fig. 2):

- **Heap** *N*: *N* deposits distributed evenly around the base at a distance D={7,9,11,13}D={7,9,11,13} m from the base edge. These deposits represent heaps of resource that have large volumes and occupy a large area, withrD=0.5rD=0.5 m and rC=3rC=3 m. For example, a Heap2 scenario contains two deposits with volume of 50 units each.

- **Scatter** *N*: *N* deposits randomly distributed between (D−5)(D−5) m and (D+5)(D+5) m from the base edge. These deposits are small, with rD=0.1rD=0.1 m and rC=1rC=1 m, and often numerous, containing a small *V*each. For example, a Scatter25 scenario contains 25 deposits with volume of four units each.

- A laden robot returning to the base deposits its load in the unloading area in the form of NpNp number of pellets of size 0.1 m33, where Np=L×(4\Lmax)Np=L×(4\Lmax), set to the closest larger integer. For example, when Lmax=1Lmax=1 and L=1L=1, a robot deposits four pellets. When Lmax=1Lmax=1 and L=0.1L=0.1, the robot deposits one pellet, etc. These pellets have the potential to cause congestion in the unloading area if many are deposited at the same time. Deposited pellets disappear from the simulation after a period of *unloading area handling time*, tHtH, representing their use or consumption by a hypothetical unmodelled system of robots or human users. By setting tH=0tH=0 s, the system is able to represent pellets that cause no congestion in the unloading area. Scenarios where tH>0tH>0 s are explored in Sect. 8.2.

3.2 Robots

The simulated marXbots (Bonani et al. 2010) are circular, differentially steered, robots with a diameter of 0.17 m that in our simulation can reach a maximum speed of 5 cm/s. They are equipped with four colour sensors pointed to the ground, a ring of 24 infrared proximity sensors used for collision avoidance, a light sensor used for navigation towards the base, a range and bearing module used for localisation of other robots and for communication, wheel-mounted sensors utilised for odometry and a ring of eight colour LEDs used for debugging. The maximum range at which the robots can detect objects via the proximity sensors is 0.3 m. Their maximum communication range via the range and bearing module is 5 m. The range and bearing module is based on line of sight, and hence intermittent ranging and communication problems are possible, although they do not have a significant effect in the scenarios we explore here. It is assumed in the simulation that the light sensor can detect the light above the base from anywhere in the experimental arena (as in, for example, Labella et al. 2004; Gutiérrez et al. 2010). Sensor noise and wheel slippage are not modelled. Consequently, there is no odometry error in our simulation.

The robot control algorithm is inspired by foraging bees, and the robots are modelled as finite state machines (Fig. 3). A robot starts in a random orientation and position on the dance floor as an *observer*, ready to receive recruitment signals. Observers move randomly on the dance floor and avoid travelling into the unloading area. When a recruitment signal is received, the robot becomes a *forager* and navigates towards a deposit location obtained from the recruiter. Alternatively, an observer still on the dance floor can become a *scout* with scouting probability p(S)=10−3 at each time step.

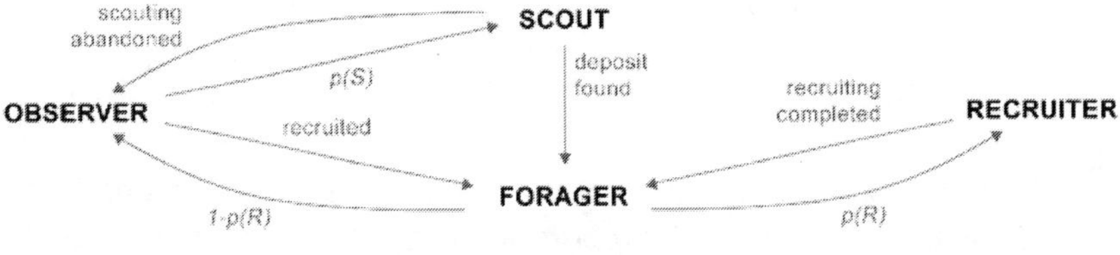

Figure 3. Finite state machine representation of the robot controller

Scouts leave the base and use Lévy movement (Reynolds and Rhodes 2009) to search for a deposit within 20 m from the base. Any scout that cannot find a deposit within 600 s returns to the dance floor and becomes an observer. While outside the base, a robot updates its estimation of the relative position of the base using path integration based on odometry at each time step (e.g. Borenstein 1998; Lemmens et al.2008; Gutiérrez et al. 2010). When a scout discovers a deposit, it becomes a forager. All foragers load *L*units of volume of the resource and determine an estimate of the deposit's energy efficiency (after Seeley1994):

$$E_E = \frac{V_i' \times Q}{d_i} \tag{1}$$

where V'i is the volume left in the deposit after the robot's visit, QiQi is the deposit quality, and didi is the odometry-estimated linear distance from the unloading area to the deposit. The robot then returns back to the base utilising phototaxis and keeps track of its relative position to the deposit using odometry.

After a forager unloads its cargo in the unloading area, it moves to the dance floor and becomes a *recruiter*with a recruitment probability *p*(*R*):

$$p(R) = \begin{cases} 1.0 & \text{if } E_E > 0 \\ 0.0 & \text{else} \end{cases} \tag{2}$$

and performs recruitment inspired by bee waggle dancing for TR=120TR=120 s, randomly moving across the dance floor and advertising its deposit location to all observers located within communication range dCdC. In order to minimise the influence of the particular direction from which it arrived, and thereby give it the chance to influence more observers, the recruiter travels to the middle of the dance floor before it starts recruiting. Similarly, forager bees enter the dance floor from a single direction through a small nest entrance (Seeley and Morse 1976), allowing potential recruits to interact with any recruiter regardless of its previous foraging location.

The deposit location is communicated to each observer in a one-to-one fashion, where local axes of the robots and their alignment relative to each other are taken into account when conveying the positional information (Gutiérrez et al. 2010). While this technique is sensitive to the accuracy of a robot's sensors (Gutiérrez et al. 2010; Miletitch et al. 2013), it removes the need for a shared reference point. Note that bees, on the other hand, orient their dances relatively to the sun (von Frisch 1967).

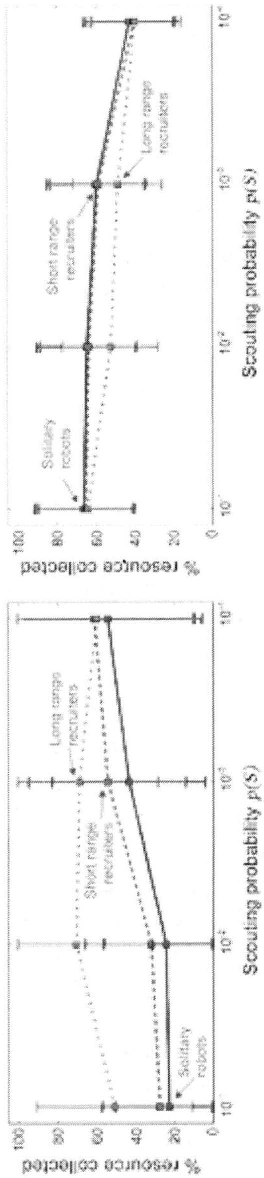

Figure 4. Foraging performance of 25-robot swarms in the **a** Heap1 and **b** Scatter25 scenarios using various*p*(*S*) values and dC=0dC=0 m (*solid line*), dC=0.6dC=0.6 m (*dashed line*) and dC=5dC=5 m (*dotted line*). Each point represents mean percentage of available resource that was collected in a given scenario, collated over 50 one-hour-long runs for each of the deposit distances in the set D={7,9,11,13}D={7,9,11,13} m. The whiskers represent 95 % confidence intervals

The recruiter always resumes foraging from the same deposit after it completes recruitment. In cases when there is no deposit to return to (p(R)=0p(R)=0), a robot does not recruit or forage again but becomes an observer instead.

If a forager reaches a deposit location but the deposit cannot be found, the robot performs neighbourhood search that lasts 180 simulated seconds and during which it moves randomly in a circular area with a radius of 2 m around the expected deposit location. If the search is unsuccessful, the robot returns to the base to become an observer. Any unsuccessful foragers and scouts are opportunistic and start foraging from a deposit if they find one on their way back to the base.

4. FORAGING PERFORMANCE IN STATIC ENVIRONMENTS

Before exploring scenarios that require plasticity, experiments were performed in Heap1, Heap2, Heap4, Scatter10 and Scatter25 scenarios that lasted T=1T=1 simulated hour and where all deposits had a constant quality Q=1Q=1 throughout the simulation. We aimed to understand how the communication range, dCdC, and the scouting probability, *p*(*S*), of robots affected the total amount of resource collected in these scenarios. Each experiment was repeated 50 times. By varying dCdC and *p*(*S*) to define three types of robot, we could encourage three types of homogeneous robot swarm: (Fig. 4):

- Solitary robots, where observers left the dance floor almost instantaneously to scout (p(S)=10−1p(S)=10−1), and robots could never recruit each other (dC=0dC=0 m)

- Short-range recruiters, where observers spent longer on the dance floor (p(S)=10−3p(S)=10−3), but robots could only recruit observers that were near to them on the dance floor (dC=0.6dC=0.6 m)

- Long-range recruiters, where robots could recruit any observer on the dance floor (dC=5dC=5 m, p(S)=10−3p(S)=10−3)

In the most extreme Heap scenario, Heap1, with only a single large deposit, using large dCdC and small $p(S)$ was the most beneficial as it allowed for mass recruitment once the deposit was discovered and foragers returned to the base to convey the deposit location to any unsuccessful scouts, making long-range recruiters the most suitable option (Fig. 4a). In the other extreme scenario, Scatter25, where there were 25 small deposits, solitary foraging led to the best results (Fig. 4b). In this case, recruitment damaged performance because of environmental interference between robots, where the robots altered each other's foraging environment by depleting low-volume deposits and where recruitment thus led to a lot of wasteful foraging trips (see Pitonakova et al. 2014 for discussion of robot–robot interference during foraging). Short-range recruiters could achieve a balance between exploration and exploitation, which made them more robust across scenarios, although they never achieved performance better than that of the other two swarm types (see the dashed line in Fig. 4).

The results shown in Fig. 4 were obtained using swarms of 25 robots. Similar relationships between dCdC and$p(S)$ were found for swarms of 15 and 35 robots. The findings also held for less extreme environments such as Heap2, Heap4 or Scatter10, although results varied between runs more in these intermediate scenarios.

5. HETEROGENEOUS DEPOSIT QUALITY

Having established that recruitment is most beneficial in Heap scenarios, where deposits are large but hard to find, we now explore what type of information transfer would allow the robots to collectively choose a deposit of a better energy efficiency. We examined the following scenarios using deposit distances $D=\{7,9,11,13\}D=\{7,9,11,13\}$ m:

- **Heap2A:** two deposits with volume $V=50V=50$ each and with qualities $Q1=0.5Q1=0.5$, $Q2=1.5Q2=1.5$
- **Heap2B:** two deposits with $V=50V=50$ each and with qualities $Q1=0.1Q1=0.1$, $Q2=1.9Q2=1.9$
- **Heap4A:** four deposits with $V=25V=25$ each and with qualities $Q1–3=0.5Q1–3=0.5$, $Q4=2.5Q4=2.5$

- **Heap4B:** four deposits with $V=25$ each and with qualities $Q1=0.1$, $Q2=0.5$, $Q3=1.5$, $Q4=1.9$

The maximum volume L_{max} loaded by a robot per foraging trip was decreased from 1.0 to 0.25 in all experiments reported below. The robots still collected $L_{max} \times Q$ units of net resource value per foraging trip and deposited a maximum of four pellets in the unloading area. Decreasing L_{max} causes swarms to carry out more foraging trips during a simulation (see online supplementary material, Fig. S1), generating a larger data set which is necessary in order to analyse foraging dynamics when the environment changes over time.

The swarms used in the previous experiments were unable to discriminate between deposits based on energy efficiency as robots always followed the first recruiter that they met and remained foraging from a deposit until it was fully depleted. On average, such swarms thus distributed themselves to forage from all deposits equally. They were used as control swarms in the following experiments.

In order to allow robots to preferentially concentrate their foraging effort on deposits with higher energy efficiency, we created swarms of ***beggers***, where all robot states and state transitions remained the same, but where recruiters asked each other about the energy efficiency of advertised deposits. If another robot's deposit had higher EE, a recruiter switched to advertise this deposit and then foraged from it. This strategy was inspired by the begging behaviour of bees described in Sect. 2.1 1. As before, we tested beggers with both short-range (0.6 m) and long-range (5 m) communication signals.

Swarms of short-range beggers were able to identify and concentrate on better deposits within the first 20 min and could thus collect more resource than the control swarms (Fig. 5a). Long-range beggers switched to better deposits faster, making them more efficient at the task, especially for larger deposit distances (Fig. 5b). The size of the difference between deposit qualities did not affect the time it took to switch to a better deposit given a particular number of deposits. However, performance of both swarms was better in the Heap2B scenarios compared to Heap2A, as the difference between deposit qualities was larger in Heap2B and it was thus more advantageous to forage from the better deposit. Similarly, when robots could concentrate on a deposit with quality 2.5 in Heap4A, they performed better than in Heap4B, where the best deposit had quality of only 1.9.

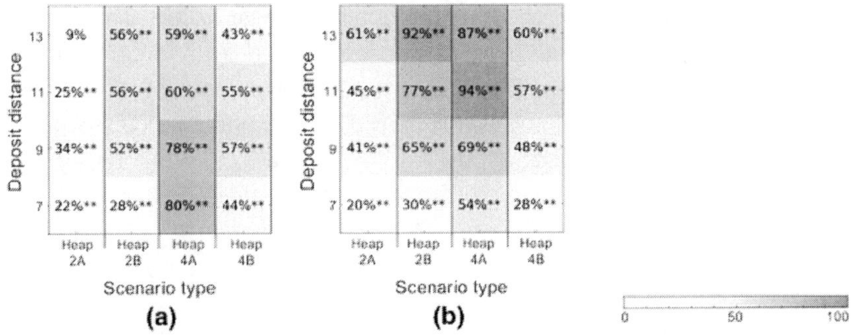

Figure 5. Difference in the amount of resource collected compared to control (non-switching) swarms using 25 beggers with **a** short-range, **b** long-range signals during one simulated hour. The measured values represent averages from 50 independent runs. Statistically significant differences are indicated with*asterisks* (Wilcoxon signed-rank test, ∗∗=p<0.01∗∗=p<0.01, ∗=p<0.05∗=p<0.05)

6. THE ABILITY OF ROBOTS TO REPEATEDLY SWITCH BETWEEN DEPOSITS

In biological settings such as nectar foraging, the quality of foraging sites changes over time. In robotic applications, a preference for where to forage may also need to change due to the nature of the task, orders from humans, etc. It is thus important not only to be able to collectively select a better foraging site, but to be able to do so repeatedly, as the environment changes. Therefore, in dynamic environments, the ability to achieve a balance between exploitation and exploration becomes important.

To test the ability of our swarms to preferentially forage from better deposits in dynamic environments over a long period of time, we ran simulations over a number of hours and changed deposit qualities (but not their locations) at the end of each *deposit quality change interval* of length TQTQ. When the change was made, deposits were assigned new qualities from a quality set given for a particular scenario and their volumes were replenished, so that the amount of available resource was the same at the beginning of each quality change interval.

In this task, strong commitment to a single best deposit could be a serious problem, as it might cause the swarm to loose track of other deposits and of new places to forage from when the quality change occurs. This was the case in all Heap2 scenarios, where beggers locked into foraging from a single deposit (Fig. 6a), an effect that was more pronounced for long-range beggers, and that held for both short (TQ=1TQ=1 h) and long (TQ=2TQ=2 h) quality change intervals.

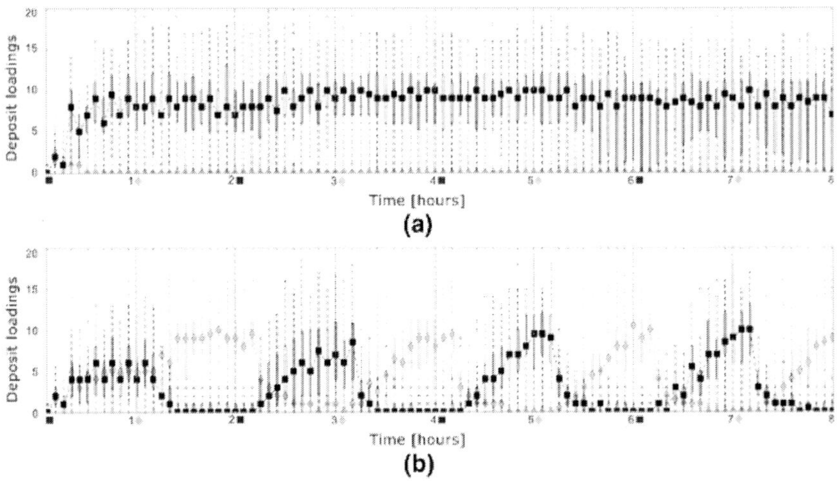

Figure 6. Median number of loadings from two deposits (*square* and *diamond symbols*) during the first 8 h of simulation using **a** 25 short-range beggers and **b** 25 short-range checkers in the Heap2B scenario with D=9D=9 m, TQ=1TQ=1 h. Deposit qualities were exchanged every hour during the experiment. The symbol of the deposit with higher quality is shown along the time axis at the beginning of each hour. Each data point represents a median value for a particular time interval and is based on a set of results collected from 50 independent runs. A data point is surrounded by a *box*, representing the inter-quartile range or "middle fifty" of the result set, and whiskers representing the "middle 97", with outliers outside this range shown as *plus signs*

On the other hand, in all Heap4 scenarios, where deposits had only half of the volume compared to the Heap2 scenarios and the differences between their energy efficiencies were thus smaller, the resulting foraging pattern depended on the communication range of robots and the length of the quality change interval used (TQTQ). Since beggers usually rapidly recruited each other to the best deposit found, large groups of robots quickly exploited a single foraging site until its energy efficiency fell below that of alternative deposits, at which point the robots rapidly switched to a different deposit. This repeated every time a certain volume of a deposit that was currently being exploited by a majority of robots was depleted, preventing the swarms from preferentially foraging from deposits with superior quality and, on average, to forage from all deposits with an equal probability (Fig. 7a). Unlike in Heap2, some members of the swarm always foraged from inferior deposits and the swarm thus retained the memory of all deposits in the environment. However, the inability to forage from deposits of superior quality for sufficiently long caused short-range beggers when TQ=1TQ=1 h and long-range beggers when TQ=1TQ=1 h and 2 h to collect less resource that the control swarms (Figs. 8a, b, 9a, b). On the other hand, when TQ=2TQ=2 h, deposit quality remained stable long enough, allowing short-range beggers to eventually choose the best deposit after a period of indecisiveness and more importantly to repeat this pattern during each quality change interval. This led to foraging performance better than that of the control swarms in both Heap4A and Heap4B scenarios when deposits were near the base (Fig. 9a).

These results suggested that a swarm behaviour where the spread of social information is regulated could potentially lead to more effective plastic reconfiguration and consequently more sustainable preferential foraging. To test this hypothesis, we introduced a modified behaviour that explicitly regulated the spread of social information. A swarm of ***checkers*** was implemented, where the robots did not compare deposit energy efficiency directly with other recruiters, as it was the case with beggers. However, instead of having a binary recruitment probability, *p*(*R*) (Eq. 2), and a fixed recruitment time, TRTR, checkers calculated these variables based on how the energy efficiency of a deposit had changed since the last time they had visited it:

$$\delta = \min\left(1.0, \frac{E_E}{E'_E}\delta'\right) \tag{3}$$

where E′ and δ′ were the energy efficiency of the deposit and δ measured during the last loading event,

$$p(R) = \delta \tag{4}$$

$$T_R = \delta \times 120\,s \tag{5}$$

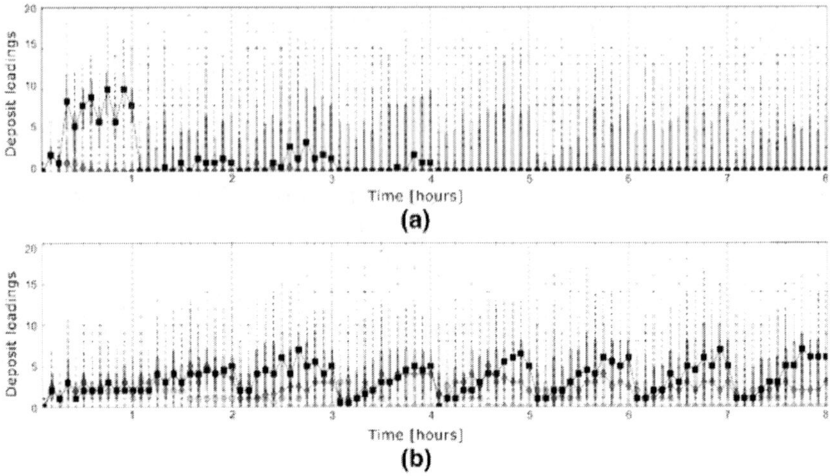

Figure 7. Median number of loadings from four deposits during the first 8 h of simulation using **a** 25 short-range beggers and **b** 25 short-range checkers in the Heap4B scenario with D=9D=9 m, TQ=1TQ=1 h, based on 50 independent runs. Deposit qualities were assigned randomly every hour during the experiment. Therefore, instead of identifying loadings from a particular deposit in a particular location, like in Fig. 6,*each symbol* identifies loadings from a deposit of a particular quality in a given time interval: Q=1.9Q=1.9(*squares*), Q=1.5Q=1.5 (*diamonds*), Q=0.5Q=0.5 (*circles*), Q=0.1Q=0.1 (*crosses*)

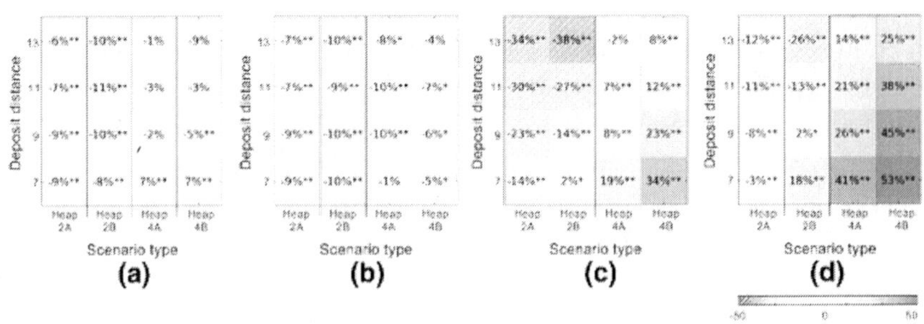

Figure 8. Difference in the amount of resource collected compared to control (non-switching) swarms using **a** 25 short-range beggers, **b** 25 long-range beggers, **c** 25 short-range checkers, **d** 25 long-range checkers during 1 h<T<131 h<T<13 h of 50 independent runs, using TQ=1TQ=1 h. Statistically significant differences are indicated with *asterisks* (Wilcoxon signed-rank test, ∗∗=p<0.01∗∗=p<0.01, ∗=p<0.05∗=p<0.05)

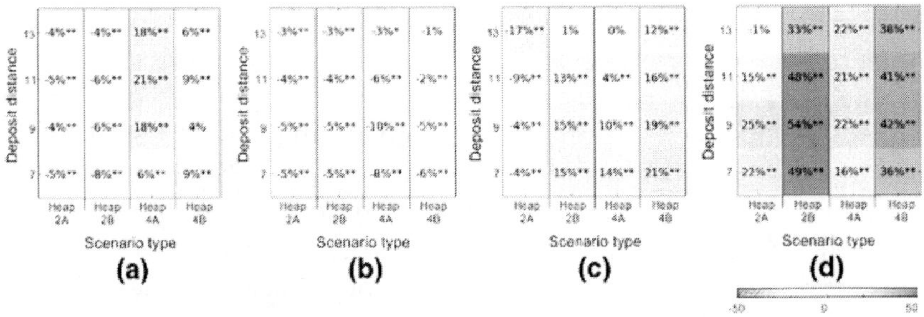

Figure 9 Difference in the amount of resource collected compared to control (non-switching) swarms using **a** 25 short-range beggers, **b** 25 long-range beggers, **c** 25 short-range checkers, **d** 25 long-range checkers during 2 h<T<262 h<T<26 h of 50 independent runs, using TQ=2TQ=2 h. Statistically significant improvements are indicated with *asterisks* (Wilcoxon signed-rank test, ∗∗=p<0.01∗∗=p<0.01, ∗=p<0.05∗=p<0.05)

Because their recruitment probability, and consequently their probability of returning to a deposit, decreased gradually as the deposit became depleted, checkers abandoned a foraging site one by one over time, allowing them to be recruited by other robots or to go scouting. By contrast, both the control swarms and beggers remained foraging from the same deposit until it became completely depleted or until, in the case of beggers, the robots received social information about a better deposit. Perhaps more importantly, the fact that checkers evaluated the change in energy efficiency of a deposit each time they visited it, enabled them to immediately abandon a deposit if its quality suddenly decreased, without having to rely on receiving knowledge from other robots about better deposits, as the beggers did.

The checker behavioural strategy was also inspired by bees, in particular by their food patch abandonment behaviour (Seeley 1994; Biesmeijer and De Vries 2001). An agent-based model, where recruitment time was influenced by quality of an advertised site but where agents, similarly to our beggers, could also switch to another site after recruitment was explored by Valentini et al. (2014).

The checker strategy allowed the swarms to repeatedly switch to a better deposit in both Heap2 (Fig. 6b) and Heap4 (Fig. 7b) scenarios, unlike the beggers that either foraged from one deposit throughout the whole run or were often unable to make a clear collective decision. The checker swarms were able to concentrate on exploiting better deposits faster when deposit distance was smaller (see online supplementary material, Figs. S2 and S3), as deposit visits, as well as recruitment, occurred more frequently (for example in Heap2B, the swarms switched to a better deposit after about 25 min when D=7D=7 m and after about 35 min when D=13D=13 m). However, note that checker swarms were unable to concentrate on the better deposit before the first quality change occurred, as a majority of robots had to abandon the deposit from which they initially foraged before recruitment to a better foraging site could take place. Such a significant abandonment of a deposit could only occur if the deposit was nearly depleted, which did not occur unless the whole swarm foraged from it, or when deposit qualities were changed suddenly by the simulation engine.

Nevertheless, the long-term behaviour of the checkers was much more desirable than that of the beggers and could significantly outperform the control swarms, especially when four deposits were placed in the environment, when the deposit quality change interval was long, or when long-range recruitment signals were used (Figs. 8c, d, 9c, d). However, in some scenarios, the checkers collected less resource than the control swarms. This was the case in all Heap2 scenarios when TQ=1TQ=1 h for both short- and long-range checkers and in the Heap2A scenario when TQ=2TQ=2 h for short-range checkers. The poor performance was caused by the fact that the checkers gradually abandoned a deposit of a worse quality after deposit qualities were changed, which led to a decrease in foraging activity. The robots that stopped foraging waited on the dance floor to be recruited or left the base to become scouts, creating a period of time when the swarm could not do any foraging work. Even though the robots eventually did forage from a deposit of a better quality, they often did not have enough time to make up for the loss of foraging time. The period of low foraging activity was especially long for short-range checkers and when the deposit distance was large. For example, it took the swarms up to 35 min to start foraging from the better deposit in the Heap2B scenario when D=13D=13 m and TQ=1TQ=1 h (see online supplementary material, Fig. S2), leading to 38 % decrease in the total amount of resource collected compared to the control swarm (Fig. 8c).

7. ANALYSING THE VALUE OF INFORMATION

It is clear that foraging performance in dynamic environments is closely related to how information is obtained and exchanged between robots.

While sudden bursts of information transfer can lead to over-commitment to a single deposit, which reduces the chance of acquiring information on alternative deposit sites, a mix of scouting and more controlled information transfer can lead to plasticity. In order to characterise behavioural strategies that are beneficial, it would be useful to have a measure that could quantify the effect of communication.

It is possible to mathematically model swarms using differential equations and then use the model to test simulation parameters and predict the average result (Lerman et al. 2006; Liu and Winfield 2010). It has also been shown that decision-making swarms can be modelled as dynamical systems that exhibit Hopf bifurcations and limit cycles (Pais et al. 2012). Finally, information theoretical measures, such as transfer entropy and local data storage, can be used to show how communication affects the state of an agent based on the states of the other agents it shares information with (Miller et al. 2014).

While mathematical models are often tractable and can generalise well (Martinoli et al. 2004), they are also often complex and difficult to apply in real-life scenarios (McFarland and Spier 1997) where many options exist and where physical interactions substantially influence the resulting swarm behaviour. Inspired by information theoretical approaches, we seek to establish a simple measure, the information value *I*, that works directly from experimental data and represents a quantifiable effect of information transfer between the environment and the swarm and within the swarm itself. This measure should reflect the flow of new information though the system, should be positive when recruitment leads to exploitation of relatively good deposits and should be negative when recruitment causes the system to lower its foraging performance.

The information value IrIr of a robot is defined as

$$I_r = Q_{\text{new}} - Q_{\text{old}} \tag{6}$$

where Qnew is the quality of a new deposit the robot finds out about and Qold is the quality of a deposit it previously foraged from. Note that Qold=0 when a robot has not foraged before. IrIr is thus equal to Qnew when a scout discovers a new deposit or when a robot that does not posses any information is recruited. In cases when a robot currently foraging from a deposit switches to another, IrIr can be either positive or negative. For example, if a robot previously foraged from a deposit with Qi=1.0Q and was recruited to a new deposit with Qi=0.5Q, Ir=−0.5. Finally, Ir=0 for all robots that do not receive any information in a given time step.

We compare deposit qualities, rather than energy efficiencies, i.e. we do not take the volume left in a deposit or its distance from the base into account. The information value thus only captures utility obtained from a single foraging trip that the robot is guaranteed to make, rather than that of possible future trips. Furthermore, since deposits in our Heap scenarios have the same distance from the base, comparing their qualities is sufficient to identify which one is more profitable.

We obtain the swarm information value *I* normalised per robot as:

$$I = \sum_{r=1}^{N_R} I_r \times \frac{1}{N_R} \qquad (7)$$

where NRNR is the total number of robots. A time series of the swarm (Fig. 10) is obtained by sampling the average value of *I* over short time intervals (in our case 300-s intervals). When a swarm working in a dynamic environment is able to repeatedly switch its attention to the current best deposit, the shape of the time series repeats for each quality change interval and has three distinguishable information value regions:

1) **A negative region** that is caused by recruitment to the previously exploited deposit(s) that have become worse since the change in deposit quality.

2) **A flat region** where new and old information coexists in the system. We refer to the negative region and the flat region together as a non-positive region.

3) **A positive region** that represents recruitment to the newly discovered better-quality deposit(s).

The size (i.e. the area between the curve and the abscissa) of negative regions represents a lag between changes in the environment and changes in the swarm's collective knowledge. This region is usually larger for long-range recruiters and for shorter deposit distances, i.e. when recruitment is stronger (see online supplementary material, Figs. S4–S6). Stronger recruitment tends to result in a larger number of robots to be foraging from the better deposit at the end of a change interval. Immediately after a quality change, there is thus a higher probability of an observer being recruited by a robot that is signalling out-of-date information.

The size of positive regions is influenced by the number of observers that a recruiter holding new information can reach, i.e. by the speed of information transfer through the swarm. It is also larger for long-range recruiters and shorter deposit distances.

The length of the non-positive regions is a combination of the positive impact of information speed and the negative impact of commitment to old information.

When robots deplete and abandon the better deposit(s) before the quality change occurs, as was the case for checkers in Heap2 scenarios when TQ=2TQ=2 h, there is no recruitment to a low-quality deposit after the quality change. Therefore, there are no negative regions of *I* and non-positive regions become very short. On the other hand, when the whole swarm commits to a single deposit and is unable to abandon it throughout the experiment, as was usually the case with long-range beggers, only a single large positive region exists at the beginning of the simulation, followed by zero information value throughout the rest of the run. Using information value, we can thus distinguish three work modes of foraging swarms (Fig. 10):

1) ***Switching*** swarms that alternate between non-positive and positive regions and sometimes also show negative regions. These swarms are able to concentrate on deposit(s) of better quality in each quality change interval.

2) ***Locked*** swarms that show a large positive region at the beginning of the simulation, but are unable to alter their chosen foraging site for the rest of the run and thus show zero communication value thereafter.

3) ***Indecisive*** swarms that also have a large positive region at the beginning, but are able to gain new information through scouting or because recruitment does not affect as many robots as in locked swarms. In contrast with switching swarms, rapid information transfer in indecisive swarms prevents the collective from settling on one solution, leading to oscillations around I=0I=0. When multiple runs are considered, time series of *I* has medians of zero and many outliers..

When runs that last several hours and have multiple quality change intervals are considered, locked and indecisive swarms always collect less resource than the control swarms that forage from all deposits with similar proportions. The relative performance of switching swarms depends on a number of factors. As we showed in the previous sections, switching to better deposits can sometimes occur too late or the total amount of loadings can be insufficient, which leads to worse performance. This was, for example, the case for checkers in Heap2 scenarios when TQ=1TQ=1 h.

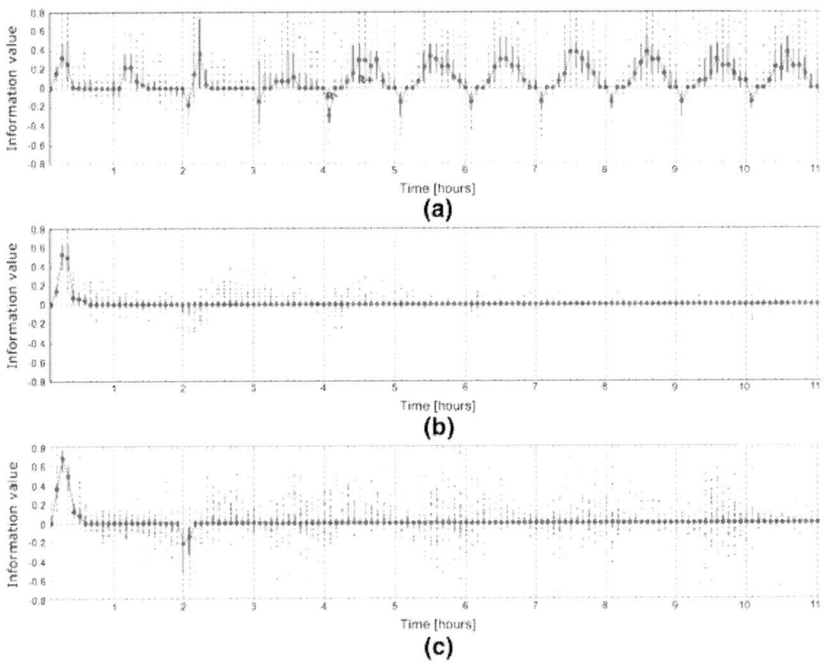

Figure 10. An information value time series based on 50 independent runs with 25 robots for **a** switching swarm: long-range checkers in Heap2B, D=9D=9 m, TQ=1TQ=1 h **b** locked swarm: long-range beggers in Heap2A, D=9D=9 m, TQ=2TQ=2 h **c** indecisive swarm: long-range beggers in Heap4B, D=9D=9 m, TQ=2TQ=2 h. Negative regions (R−−) and positive regions (R+) are shown for the switching swarm

In the following sections, we test our communication strategies further under various experimental perturbations and use the information value measure, *I*, to analyse the results. We show that the switching work mode is more common when information transfer is slower and that measuring *I* can thus help robot designers to select behavioural parameters that lead to effective plastic self-organisation.

8. VARYING EXPERIMENTAL PARAMETERS

It is important to understand how the chosen experimental parameters affect the performance and work modes of our swarms. In this section, we explore the influence of the shape of the dance floor, of the tendency for collected pellets to accumulate and interfere with robot movement, and of the size of the robot swarms. We evaluate how varying these aspects of the foraging scenario affects the behaviour of all four types of swarms (i.e. short-range and long-range beggers, and short-range and long-range checkers) compared to performance under normal conditions (i.e. circular dance floor, no pellet accumulation and 25-robot swarms—see Sects. 5 and 6). We also identify cases in which variation to these aspects of the scenario causes a swarm to transition from one work mode to another (e.g. from switching to indecisive). See Fig. 11 for a summary of all experiments and the work modes exhibited by the swarms.

8.1 Restricted dance floor shape

In the previous experiments, a returning forager travelled to the middle of the base before it started recruiting in order to be able to interact with any observer, regardless of direction from which it returned to the base. However, the constraints of a particular foraging scenario might require a different approach. For example, we might want to create a recharging area in a part of the recruitment area or otherwise change the shape of the base and thus indirectly alter the way in which robots meet each other and communicate. Information flow throughout a swarm will tend to be influenced by such physical logistics, potentially changing the ability of the swarm to maintain plasticity and thereby influencing performance levels.

In the following experiment, we designated an inaccessible circular area in the middle of the base, making the dance floor doughnut shaped and 35 cm thick. While the robots could still see other robots across the inaccessible central area, and could thus communicate with them if their communication range allowed for it, they could not move through the central restricted area. Consequently, while undertaking a random walk within the dance floor, a robot tended to remain relatively close to the point at which it first entered the base. The restricted shape of the dance floor also resulted in higher congestion and constrained movement of the robots within and immediately around the dance floor. The congestion was more severe when long-range recruitment was used or when deposits were closer to the base, i.e. when the number of foragers returning to the base at the same time was high. Additionally, the annular shape of the dance floor restricted communication within swarms with short-range recruitment, as it made interactions between robots that foraged from different areas of the environment less probable and prevented information from spreading as easily throughout the swarm.

<table>
<tr><th colspan="2"></th><th>Short-range beggers</th><th>Long-range beggers</th><th>Short-range checkers</th><th>Long-range checkers</th></tr>
<tr><td rowspan="4">Normal conditions</td><td>Heap2, TQ = 1h</td><td rowspan="2">Locked</td><td rowspan="2">Locked</td><td rowspan="4">Switching</td><td rowspan="4">Switching</td></tr>
<tr><td>Heap2, TQ = 2h</td></tr>
<tr><td>Heap4, TQ = 1h</td><td>Indecisive</td><td rowspan="2">Indecisive</td></tr>
<tr><td>Heap4, TQ = 2h</td><td>Switching</td></tr>
</table>

<table>
<tr><th colspan="2"></th><th>Short-range beggers</th><th>Long-range beggers</th><th>Short-range checkers</th><th>Long-range checkers</th></tr>
<tr><td rowspan="4">Restricted dance floor shape</td><td>Heap2, TQ = 1h</td><td rowspan="3">Delayed switching</td><td rowspan="2">Locked</td><td rowspan="4">Switching</td><td rowspan="4">Switching</td></tr>
<tr><td>Heap2, TQ = 2h</td></tr>
<tr><td>Heap4, TQ = 1h</td><td rowspan="2">Indecisive</td></tr>
<tr><td>Heap4, TQ = 2h</td><td>Switching</td></tr>
</table>

<table>
<tr><th colspan="2"></th><th>Short-range beggers</th><th>Long-range beggers</th><th>Short-range checkers</th><th>Long-range checkers</th></tr>
<tr><td rowspan="4">Increased pellet accumulation</td><td>Heap2, TQ = 1h</td><td rowspan="2">Locked</td><td rowspan="2">Locked</td><td rowspan="2">Switching</td><td rowspan="2">Switching</td></tr>
<tr><td>Heap2, TQ = 2h</td></tr>
<tr><td>Heap4, TQ = 1h</td><td>Indecisive</td><td rowspan="2">Indecisive</td><td rowspan="2">Indecisive</td><td rowspan="2">Indecisive</td></tr>
<tr><td>Heap4, TQ = 2h</td><td>Indecisive</td></tr>
</table>

<table>
<tr><th colspan="2"></th><th>Short-range beggers</th><th>Long-range beggers</th><th>Short-range checkers</th><th>Long-range checkers</th></tr>
<tr><td rowspan="4">Increased swarm size</td><td>Heap2, TQ = 1h</td><td>Delayed switching</td><td rowspan="2">Delayed switching</td><td rowspan="4">Switching</td><td>Delayed switching</td></tr>
<tr><td>Heap2, TQ = 2h</td><td rowspan="2">Switching</td><td rowspan="3">Switching</td></tr>
<tr><td>Heap4, TQ = 1h</td><td rowspan="2">Switching</td></tr>
<tr><td>Heap4, TQ = 2h</td><td>Switching</td></tr>
</table>

Figure 11. Work modes exhibited by the swarms under various experimental set-ups. Work modes that differ from those exhibited under normal conditions (see text) are shown in bold

Although most swarms did not change their work modes, information value analysis showed that restricting dance floor shape decreased the rate at which new information about the best deposit(s) spread through the swarm (see online supplementary material, Figs. S7, S8). For example, the total size of positive regions (i.e. sum of all I>0I>0) was 40 % smaller for short-range checkers in Heap2s (D=9D=9 m and TQ=1TQ=1 h) compared to normal conditions. The size of positive regions was less affected for long-range checkers (only 28 % smaller), but higher congestion caused the positive regions to appear much later than under normal conditions (usually after 35 instead of 25 min).

Short-range beggers were the only swarm that exhibited different work modes compared to the normal conditions (Fig. 11). The decreased probability of interactions between recruiters and observers on the dance floor caused the swarms to escape the locked mode they experienced with a full dance floor in Heap2 scenarios, allowing them to preferentially forage from better deposits. However, clear positive regions of information value were absent while negative regions remained, meaning that the information about new better deposits spread too slowly to be of use. We term this a *delayed switching* work mode (Fig. 12). The swarms also changed their work mode from indecisive to delayed switching in Heap4 scenarios with TQ=1TQ=1 h, as constrained information transfer prevented the confusion that had previously resulted from strong competition between deposits with equivalent energy efficiencies. Despite the fact that, when in the delayed switching mode, short-range beggers were able to alternate between deposits as the deposit quality changed, the foraging performance of the swarms deteriorated compared to normal conditions (Figs. 13a, 14a). First, the total number of foraging trips was smaller due to congestion caused by a smaller dance floor. Second, the delayed switching mode led to a similar number of loadings from deposits of higher and lower quality when the whole run was considered, making the proportion of loadings from the better deposit(s) similar to when the swarms were in the locked or indecisive modes (see online supplementary material, Figs. S9, S10).

The impact of congestion caused by a restrictive dance floor was most clearly observed for long-range beggers. While a recruiter's signal could reach any observer on the dance floor regardless of their relative positions and the work mode of the swarms thus remained locked, foraging performance still deteriorated as the robots spent significantly more time avoiding each other (Figs. 13b, 14b).

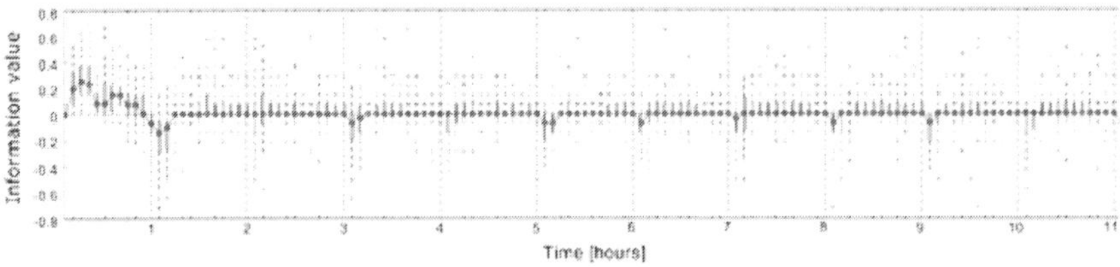

Figure 12. An information value time series based on 50 independent runs for a delayed switching swarm of 25 short-range beggers in Heap2B, D=9D=9 m, TQ=1TQ=1 h with restricted dance floor shape

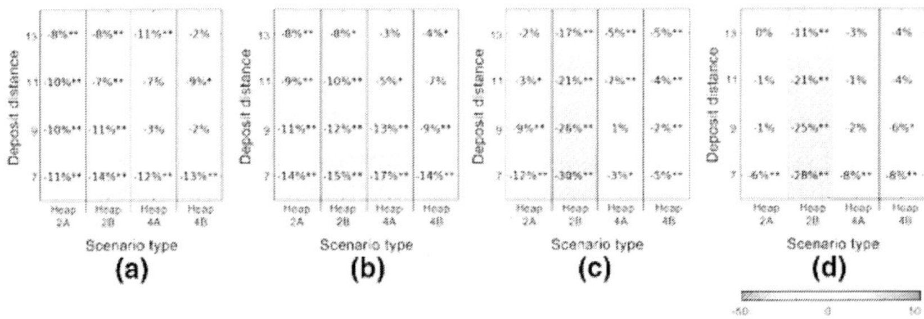

Figure 13. Difference in the amount of resource collected using an annular dance floor compared to the same communication strategy operating over a full dance floor with **a** 25 short-range beggers, **b** 25 long-range beggers, **c** 25 short-range checkers, **d** 25 long-range checkers during 1 h <T<13<T<13 h of 50 independent runs, using TQ=1 h. Statistically significant differences are indicated with *asterisks*(Wilcoxon signed-rank test, ∗∗=p<0.01∗∗=p<0.01, ∗=p<0.05∗=p<0.05)

While the restricted movement of recruiters affected both the short- and long-range checkers negatively in almost all scenarios (Figs. 13c, d, 14c, d), it had a surprisingly positive effect on short-range checkers in Heap4B scenarios when TQ=2TQ=2 h (Fig. 14c). In this case, the impaired ability of robots to recruit each other caused the swarms to forage from the best two deposits in parallel and thus to collect a total amount of resource that was higher than under normal conditions.

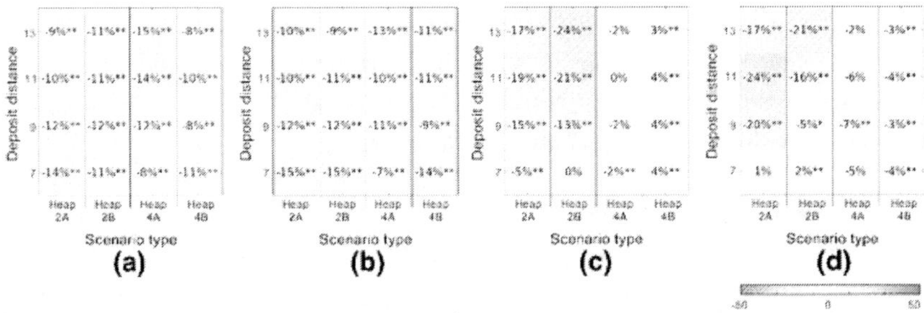

Figure14. Difference in the amount of resource collected using an annular dance floor compared to the same communication strategy operating over a full dance floor with **a** 25 short-range beggers, **b** 25 long-range beggers, **c** 25 short-range checkers, **d** 25 long-range checkers during 2 h <T<<T<26 h of 50 independent runs, using TQ=2 Statistically significant differences are indicated with *asterisks*(Wilcoxon signed-rank test, ∗∗=p<0.01∗∗=p<0.01,

8.2 Pellet accumulation

The material collected by robots in the previous experiments disappeared instantly when it was placed in the unloading bay, i.e. the unloading area handling time tH=0tH=0 s. However, in real-world applications, material would have to be stored somewhere for later use by humans or other robots. It is therefore reasonable to assume that it would accumulate and thus affect the swarm's ability to collect more. To test performance of our swarms under such conditions, we set tH={10,20}tH={10,20} s, i.e. pellets dropped by robots could accumulate in the unloading area and create congestion. Since the congestion prevented robots from foraging with maximum efficiency, and from reaching the dance floor in order to communicate, swarms always collected less resources than under normal conditions.

When pellets accumulated, both the negative and positive regions of information value were more flat, while the non-positive regions usually became longer, indicating that limited access to and from the base prevented information transfer. This effect was observed more strongly for long-range recruiters and the higher value of tHtH (see online supplementary material, Figs. S11, S12), i.e. when more foragers returned to the base at the same time and when pellets remained in the unloading area for a longer period of time. For example, when tH=20tH=20 s, the total size of positive regions (i.e. sum of all I>0I>0) in Heap2s (D=9D=9 m and

TQ=1TQ=1 h) was 67 % smaller for short-range checkers and 75 % smaller for long-range checkers compared to the normal conditions. The length of non-positive regions decreased by 25 % for short-range checkers and increased by 60 % for long-range checkers.

Since the rate at which the robots communicated was affected, the ability of swarms to switch between deposits was impaired. This caused short-range beggers, short-range checkers and long-range checkers to become indecisive in Heap4 scenarios (Fig. 11). There were no significant changes in work modes observed for long-range beggers, as they already exhibited the locked (in Heap2) and indecisive (in Heap4) work modes under normal conditions.

Pellet accumulation had also an interesting effect on the advantage of long-range over short-range recruitment. For example, under normal conditions, long-range checkers always outperformed short-range checkers in terms of the total amount of resource collected due to their ability to recruit to the best deposit faster. However, when pellets accumulated, this faster recruitment resulted in stronger congestion and in subsequent worse performance compared to short-range checkers in some scenarios (Fig. 15). This was especially true in the Heap2 scenarios when the unloading area handling time tH=20tH=20 s and when the swarms were given a long time to forage from the selected deposit (TQ=2TQ=2 h) (Fig. 15d).

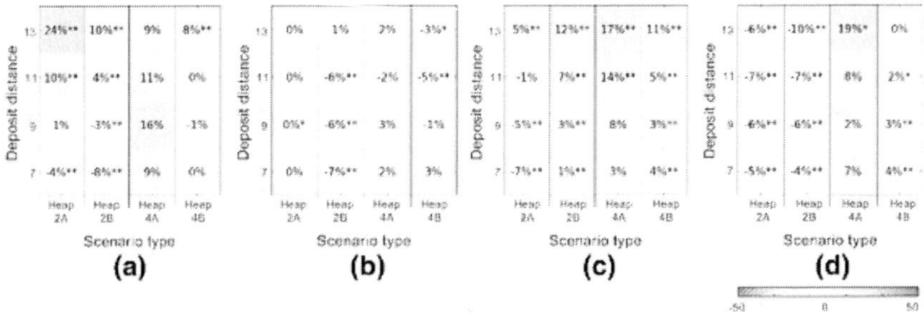

Figure 15 Improvement in resource collected using 25 long-range checkers compared to 25 short-range checkers during 50 independent runs with a unloading area handling time tH = 10 s, TQ = 1 h and 1 h < T < 13 h, b tH = 20 s, TQ = 1 h and 1 h < T < 13 h, c tH = 10 s, TQ = 2 h and 2 h < T < 26 h, d tH = 20 s, TQ = 2 h and 2 h < T < 26 h. Statistically significant improvements are indicated with asterisks (Wilcoxon signed-rank test, ∗∗ = $p < 0.01$, ∗ = $p < 0.05$)

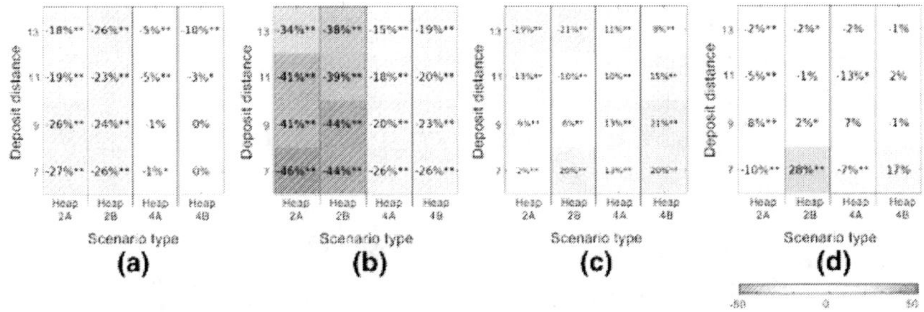

Figure 16. Improvement in resource collected compared to control (non-switching) swarms of 65 robots using a 65 short-range beggers, b 65 long-range beggers, c 65 short-range checkers, d 65 long-range checkers during 1 h < T < 13 h of 50 independent runs, TQ = 1 h. Statistically significant improvements are indicated with asterisks (Wilcoxon signed-rank test, ∗∗ = p < 0.01, ∗ = p < 0.05)

8.3 Increased swarm size

All previous experiments were performed with swarms of 25 robots. Here, we simulate swarms of 45 and 65 robots.

While short- and long-range beggers previously exhibited the locked work mode in Heap2 scenarios, higher scouting success of the swarm, caused by a higher number of robots that were scouting at the beginning of a simulation run, led to a more even distribution of robots between the deposits. This allowed larger swarms to preferentially forage from better deposits and exhibit the delayed switching work mode, or the switching work mode in the case of short-range beggers when TQ=2TQ=2 h or when NR=65NR=65 (see Fig. 11 and online supplementary material, Fig. S13). However, the significant delays in recruitment to a better deposit usually caused a stronger disadvantage for the beggers over the control swarms compared to when the number of robots NR=25NR=25 (compare Fig. 16a, b with 8a, b and 17b with 9b). In Heap4 scenarios, where the beggers were previously indecisive, the swarms were able to achieve the switching mode regardless of the value of TQTQ (Fig. 11). The switching was effective enough to allow both short- and long-range beggers to outperform the control swarms when NR=45NR=45. For example, short-range beggers collected 17 % more resources than the control swarms of the same size in Heap4A when D=7D=7 m and NR=45NR=45. However, their advantage largely disappeared when NR=65NR=65 (Figs. 16a, 17a) due to rapid exploitation of the better foraging sites and consequent ineffective foraging from other deposits, as well as due to increased congestion in the base. The disadvantage of 65-robot swarms was more pronounced for long-range beggers (Figs. 16b, 17b) that shared information with each other faster and thus exploited deposits more rapidly.

Checkers generally retained their ability to switch between deposits when the swarm size increased. The information value of the swarms, normalised per robot, did not change significantly, although the positive regions at the beginning of simulation runs were more prominent compared to when 25 robots were used (see online supplementary material, Fig. S14). This suggests that information spread faster at the beginning of a run, causing the swarms to be able to exhibit periodic switching behaviour sooner. As with beggers, checkers performed better than the control swarms when NR=45NR=45. Long-range checkers lost their advantage when NR=65NR=65 for similar reasons to the beggers (Figs. 16d, 17d). It was also observed that 65-robot swarms of long-range checkers exhibited the delayed switching work mode instead of the switching work mode in Heap2 scenarios when TQ=1TQ=1 h as a result of very significant commitment to a single deposit in each quality change interval and consequent insufficient scouting. Furthermore, there was a large variance in the total amount of collected resource across multiple runs when long-range checkers were used and when NR=65, TQ=2 h. On the other hand, large swarms of short-range checkers, where information could not spread as quickly, performed well in most of the tested scenarios (Figs. 16c, 17c). This result is not that surprising. If there is an optimal number of robots to find and exploit resources in a given environment, then swarms with long-range communication, where information spreads faster, reach that optimal number sooner and thus also become disadvantaged sooner as NRNR increases.

9. DISCUSSION

9.1 Summary of results

We have demonstrated that the nature of communication between individual robots affects a swarm's plasticity and thereby its ability to forage effectively in various types of environment. In particular, the following principles apply:

- Recruitment is most useful when deposits are large and hard to find. Conversely, recruitment can harm foraging performance when resource is distributed over numerous small deposits (see Fig. 4).

- Longer communication range increases foraging performance when deposits are large and hard to find (see Fig. 4), unless rapid exploitation of deposits can cause congestion, e.g. when a lot of the collected material accumulates in the base (see Fig. 15d) or when swarm size is large (see Figs. 16b, d, 17b, d).

- If robots need to choose foraging sites based on deposit energy efficiency in *static* foraging environments, maximising information flow in the swarm is beneficial as it maximises exploitation of the best sources (see Fig. 5).

- When environments are *dynamic* and deposit qualities change over time, both continuous exploitation and exploration of the environment are important. Regulation of information flow, for example by short communication range of robots or by an implementation of individual-based decisions on when to ignore social information, is beneficial as it prevents over-exploitation of a single foraging site (see Figs. 8c, d, 9c, d). A slower information flow thus helps to avoid situations in which a swarm locks into foraging from a single deposit or is unable to discriminate deposits of varying qualities and forages from multiple sources simultaneously.

- Inhibition of social information spread can also be achieved when robots are to some extent physically prevented from meeting and recruiting each other, for example when the shape of the base prevents robots from communicating with those that approach from the opposite side (see online supplementary material, Figs. S7, S8), or when collected material accumulates in the base (see online supplementary material, Figs. S11, S12).

- On the other hand, information is collected and spreads faster when swarm size increases (see online supplementary material, Figs. S13, S14). This can be beneficial especially when robot communication is short range (see Figs. 16c, 17a, c) or in scenarios where smaller swarms cannot alternate between deposits that change their qualities (see Fig. 11).

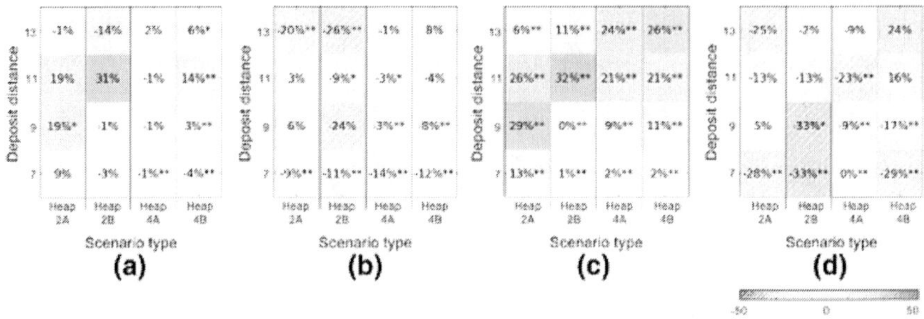

Figure 17. Improvement in resource collected compared to control (non-switching) swarms of 65 robots using a 65 short-range beggers, b 65 long-range beggers, c 65 short-range checkers, d 65 long-range checkers during 2 h < T < 26 h of 50 independent runs, TQ = 2 h. Statistically significant improvements are indicated with asterisks (Wilcoxon signed-rank test, $** = p < 0.01$, $* = p < 0.05$)

9.2 Related work

Our results are consistent with existing studies in the literature. For example, it has been shown that communication between robots is most beneficial when deposits are scarce (e.g. Liu et al. 2007; Pitonakova et al. 2014) and that short-range communication is most suitable for large robot groups, as too much information can lower task specialisation (Sarker and Dahl 2011). Previous work on the effects of communication range also demonstrated that global information exchange leads to over-commitment to a single resource (Tereshko and Loengarov 2005). Similarly, our swarms of beggers, where information spreads quickly, often found themselves committed to a single source in scenarios with two deposits.

The importance of recruitment when flower patches have a high return has been observed in honey bee colonies (Donaldson-Matasci and Dornhaus 2012). Bees, like our robots in Heap scenarios, can benefit from communicating about a location that affords many successful foraging trips. The fact that recruitment increases the probability of individuals foraging is also important in our Heap scenarios, where robots found it difficult to discover deposits. Similarly, in an agent-based model of a bee colony, recruitment was important in environments with few flower patches (Dornhaus et al. 2006).

A similar scenario has been explored by the HoFoReSim model (Schmickl et al. 2012), where simulated bees foraged from nectar sources, the quality of which changed once during a simulation run. The colony consisted of foragers that brought nectar into the nest and receivers that processed it. If a forager could not find a receiver to unload the nectar to, i.e. when the nest's nectar intake rate was too high for the receivers to handle, the forager did not waggle dance, but instead performed a tremble dance in order to activate additional dormant receivers. The authors argued that the ratio of active receivers to foragers played an important role in the ability of the colony to switch between different sources of nectar when quality changes occurred. Foragers were able to switch to a better source of nectar more quickly when there were relatively few receivers. In this situation, successful foraging overloaded receivers, causing foragers to spend time activating more receivers rather than recruiting more foragers to their foraging sites. This reduction in forager recruitment allowed bees to be less committed to high-quality foraging sites, making responses to a change in source quality faster. The role played by scarce receivers was therefore one of information regulation. In our simulation, regulation of information spread is achieved differently, e.g. by shorter communication range or constraints imposed on movement of the robots, but it has the same effect, allowing the swarms to respond to changes in the environment more appropriately. Under this reading, information flow considerations help explain a swarm's ability to exhibit plasticity both in our simulation and in HoFoReSim.

9.3 Swarm-level plasticity

In order to analyse foraging swarms, we quantified the value of information transfer and identified the following four work modes that a swarm could reach: switching, delayed switching, locked and indecisive. Swarms in the switching mode could respond to changes in deposit qualities well and usually performed better than swarms in other work modes (for example, see Figs. 8c, d, 9c, d, 9a for Heap4). Delayed switching, locked and indecisive modes were associated with worse levels of performance, although the extent of their disadvantage varied based on the environment and the swarm size.

Figure 18 shows the work modes exhibited by swarms using the different behavioural strategies explored here and the conditions under which they changed from one mode to another. The strategies are ordered from top to bottom by their ability to spread information, i.e. by the size of the positive information value region following the first deposit quality change. We tested the swarms under normal conditions, explored in Sect.6, and under three kinds of perturbation (restricted dance floor shape, pellet accumulation and increased swarm size), explored in Sect.8, giving us four foraging conditions in total. In addition, we explored four different scenario types: Heap2 versus Heap4, and TQ=1 h versus TQ=2 h. We could thus compare the behavioural strategies across a total of 4×4=164×4=16 different experimental set-ups (Fig. 11).

Short-range checkers were in the switching mode in all four scenario types under normal foraging conditions and exhibited the indecisive mode only when material accumulation increased in the Heap4 scenarios with TQ=1 h and TQ=2 h. We can say that the swarms were in the switching mode in 14 out of 16 (87.5 %) of cases. Long-range checkers also found themselves in the switching mode in all scenarios under normal conditions, but there were three conditions under which they became either indecisive or delayed. These swarms were in the switching mode in 13 out of 16 (81.25 %) cases. Moving along the information speed axis, short-range beggers were in the switching mode in only 5 out of 16 (31.25 %) cases, and long-range beggers in only 2 out of 16 (12.5 %) cases.

It is clear that rapid information spread often leads to the delayed, locked or indecisive modes, i.e. it prevents plastic self-organisation. However, it is important to point out that foraging performance in static environments was higher when information spread was fast. Furthermore, while the use of long-range recruitment in dynamic environments decreased the number of experimental set-ups where swarms were in the switching mode, it also led to a greater amount of resource being collected when the switching mode did occur and when congestion was not too high due to large swarm size or material accumulation. We could thus say that while using a robot behavioural strategy where information spreads quickly might be desirable in some application scenarios, designers of swarms have to be more certain about foraging conditions like swarm size, material accumulation, number of deposits or the length of the deposit quality change interval. On the other hand, while they were never the best in terms of foraging performance, swarms of short-range checkers, where information spread slowly, exhibited plasticity more commonly, as they retained their ability to switch between deposits under most tested conditions. Such a strategy would thus be more suitable in unknown or uncontrollable environments.

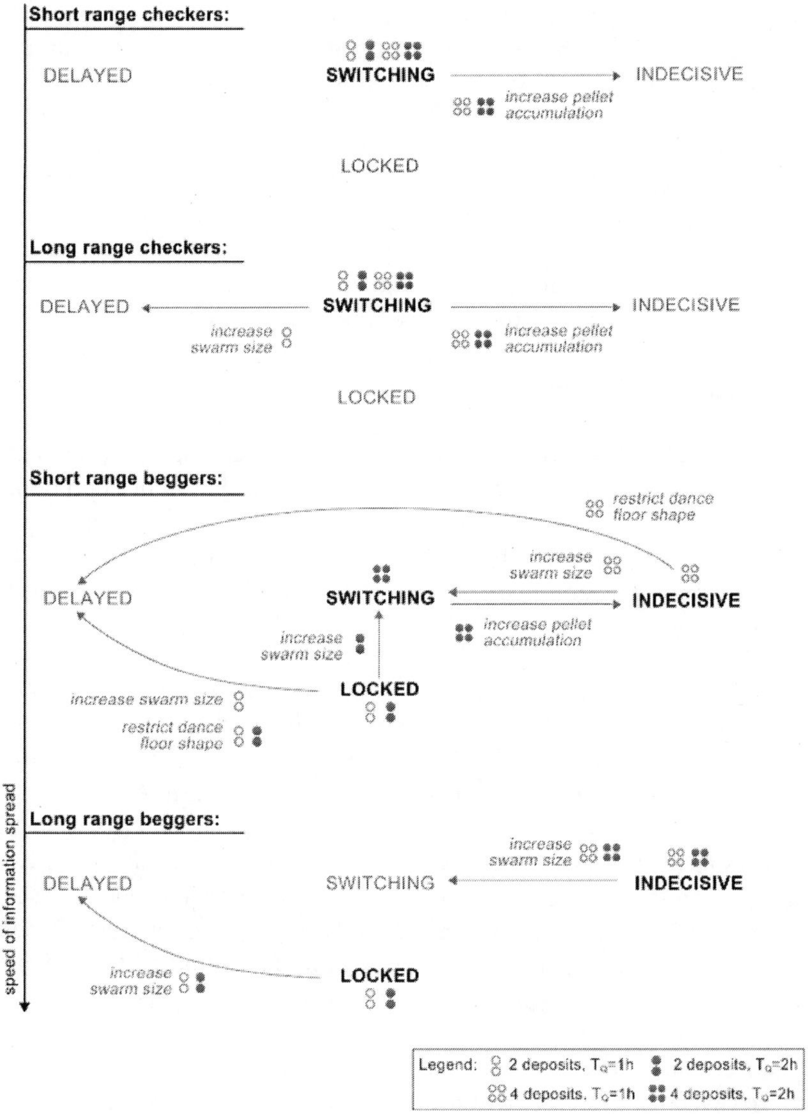

Figure 18 . Work modes and mode transitions exhibited by swarms with different behavioural strategies. The default modes that swarms operated in under normal conditions, i.e. full dance floor, no pellet accumulation, 25 robots (see Sects. 5 and 6), are shown in *bold*. Scenario markers related to a combination of the number of deposits and the length of deposit quality change interval, TQTQ, are described in the figure legend. The markers are placed above default work modes, and the transition*arrows* to indicate the scenarios in which they occurred

These findings on the effects of information flow could be applied to a wider range of problems. In particular, the following swarm robotic tasks are the most relevant:

- A number of researchers are interested in collection from different deposit types at the same time (e.g. Balch 1999; Campo and Dorigo 2007). Results here show that inhibition of social information spread, which can be achieved by reducing communication range or restricting the opportunities for communication to take place (in our study, this is effected by restricting the shape of the recruitment area or altering the character of pellet accumulation), can lead to the desired outcome of foraging from different deposits in parallel.

- Communication about resource availability was used in robots that could adapt their resting and waking thresholds based on food density (Liu and Winfield 2010). The speed of information spread, dependent on design decisions about communication type and range, could potentially play an important role in such a task and affect a swarm's flexibility.

- Task allocation has received substantial interest (e.g. Zhang et al. 2007; Sarker and Dahl 2011; Jevtic et al.2012; Zahadat et al. 2013). It is somewhat similar to deposit selection, as robots need to collectively explore the space of possible work sites and perform work on them. Principles of information transfer could be applied here, for example, by substituting deposit energy efficiency for work site importance or utility during action selection and during transmission of social information. Similarly, information value, introduced in Sect. 7, could be adapted to relate to work site importance rather than deposit quality. We aim to explore this domain in our future work.

- As an alternative to design by hand, swarm behaviour design can be achieved by reinforcement learning (e.g. Pérez-Uribe 2001) or artificial evolution (e.g. Doncieux et al. 2015; Ferrante et al. 2015; Francesca et al. 2015). Knowing what types of communication strategies are beneficial under various conditions or at least which ones are more flexible could minimise the parameter search space for these optimisation algorithms and potentially decrease complexity of the resulting behaviours or save some simulation time.

Finally, it would be possible to utilise information value in an adaptive robot control algorithm in order to make a swarm more autonomous. For example, if robots could identify pathological effects of fast or slow information transfer, they could alter their individual behaviour or their communication range in order to achieve a more effective work mode.

9.4 Conclusion

Swarm foraging has many robotic applications and can also be used as a paradigm for other swarm behaviours such as task allocation, labour division or robot dispersion. Social insects like ants and bees give an example of how powerful swarm intelligence can be and how plastic adaptive behaviour can emerge from interactions between relatively limited system parts. However, before we can use robotic swarms in the real world, we need to understand how to design individual agents for reliable collaborative work in dynamic environments.

We argued that the way in which the character of robot–robot communication influences the flow of information through a swarm and the swarm's ability to obtain new information from the environment is an important factor that affects swarm plasticity and, as a consequence, foraging performance. Behavioural strategies that limit the spread of information promote plastic self-organisation and are thus more suitable for unknown or highly dynamic environments. We have also demonstrated a method for measuring the value of information that can help us identify such plastic behavioural strategies, not only in foraging swarms but also in a wider range of swarm behaviours where social information is utilised.

FOOTNOTES

1The begging behaviour of our robots is simplified compared to that of bees. Bees integrate information received from multiple nest mates, and their foraging preferences can be affected for a number of days (De Marco and Farina 2001; Farina et al. 2005).

ACKNOWLEDGMENTS

This work was supported by an EPSRC Doctoral Training Centre grant (EP/G03690X/1). All source code and data supporting this study are openly available from the University of Southampton repository at http://dx.doi.org/10.5258/SOTON/385724 .

REFERENCES

1. Alers, S., Bloembergen, D., Hennes, D., Jong, S. D., Kaisers, M., Lemmens, N.,Tuyls, K., Weiss, G. (2011). Bee-inspired foraging in an embodied swarm. In K. Tumer, P. Yolum, L. Sonenberg, P. Stone (Eds.), *Proceedings of the 10th international conference on autonomous agents and multiagent systems (AAMAS 2011)*, (pp.1311–1312). New York: ACM.

2. Arab, A., Carollo Blanco, Y., & Costa-Leonardo, A. M. (2012). Dynamics of foraging and recruitment behavior in the asian subterranean termite coptotermes gestroi (Rhinotermitidae). *Psyche: A Journal of Entomology, 2012*, Article ID 806782.

3. Arkin, R. C. (1992). Cooperation without communication: Multiagent schema-based robot navigation. *Journal of Robotic Systems*, *9*(3), 351–364.

4. Balch, T. (1999). The impact of diversity on performance in multi-robot foraging. In O. Etzioni, J. P. Müller, J. M. Bradshaw (Eds.), *Proceedings of the third annual conference on autonomous agents*, (pp. 92–99). New York: ACM.

5. Balch, T., & Arkin, R. C. (1994). Communication in reactive multiagent robotic systems. *Autonomous Robots*,*1*(1), 27–52.

6. Biesmeijer, J. C., & De Vries, H. (2001). Exploration and exploitation of food sources by social insect colonies: A revision of the scout-recruit concept. *Behavioral Ecology and Sociobiology*, *49*(2), 89–99.

7. Bonani, M., Longchamp, V., Magnenat, S., Philippe, R., Burnier, D., Roulet, G., et al. (2010). The marXbot, a miniature mobile robot opening new perspectives for the collective-robotic research. In

Proceedings of the 2010 IEEE/RSJ international conference on intelligent robots and systems (IROS 2010), (pp. 4187–4193). Piscataway, NJ: IEEE Press.

8. Borenstein, J. (1998). Experimental results from internal odometry error correction with the OmniMate mobile robot. *IEEE Transactions on Robotics and Automation*, *14*(6), 963–969.

9. Campo, A., Dorigo, M. (2007). Efficient multi-foraging in swarm robotics. In F. Almeida e Costa (Ed.),*Proceedings of the 9th European conference on advances in artificial life (ECAL 2007)*, (pp. 696–705) Berlin: Springer.

10. Campo, A., Gutiérrez, Á., Nouyan, S., Pinciroli, C., Longchamp, V., Garnier, S., et al. (2010). Artificial pheromone for path selection by a foraging swarm of robots. *Biological Cybernetics*, *103*(5), 339–352.

11. Dai, H. (2009). Adaptive control in swarm robotic systems. *The Hilltop Review*, *3*(1), 54–67.

12. De Marco, R., & Farina, W. M. (2001). Changes in food source profitability affect the trophallactic and dance behavior of forager honeybees (*Apis mellifera* L.). *Behavioral Ecology and Sociobiology*, *50*(5), 441–449.

13. De Marco, R., & Farina, W. M. (2003). Trophallaxis in forager honeybees *Apis mellifera*: Resource uncertainty enhances begging contacts? *Journal of Comparative Physiology A*, *189*, 125–134.

14. Donaldson-Matasci, M. C., & Dornhaus, A. (2012). How habitat affects the benefits of communication in collectively foraging honey bees. *Behavioral Ecology and Sociobiology*, *66*(4), 583–592.

15. Doncieux, S., Bredeche, N., Mouret, J. B., & Eiben, A. E. (2015). Evolutionary robotics: What, why, and where to. *Frontiers in Robotics and AI*, *2*(4),

16. Dornhaus, A., Klugl, F., Oechslein, C., Puppe, F., & Chittka, L. (2006). Benefits of recruitment in honey bees: Effects of ecology and colony size in an individual-based model. *Behavioral Ecology*, *17*(3), 336–344.

17. Drogoul, A., & Ferber, J. (1993). From Tom Thumb to the Dockers: Some experiments with foraging robots. In J. Meyer, H. L. Roitblat, & S.

W. Wilson (Eds.), *From animals to animats II* (pp. 451–459). Cambridge: MIT Press.

18. Ducatelle, F., Di Caro, G. A., Pinciroli, C., & Gambardella, L. M. (2011). Self-organized cooperation between robotic swarms. *Swarm Intelligence, 5*(2), 73–96.

19. Farina, W. M., Grüter, C., & Díaz, P. C. (2005). Social learning of floral odours inside the honeybee hive.*Philosophical Transactions of the Royal Society of London, Series B, 272*, 1923–1928.

20. Ferrante, E., Duéñez Guzmán, E. (2013). GESwarm: Grammatical evolution for the automatic synthesis of collective behaviors in swarm robotics. In C. Blum C (Ed.), *Proceedings of the fifteenth international conference on genetic and evolutionary computation conference companion (GECCO 2013)*, (pp. 17–24). ACM.

21. Ferrante, E., Turgut, A. E., Duéñez-Guzmán, E., Dorigo, M., & Wenseleers, T. (2015). Evolution of self-organized task specialization in robot swarms. *PLoS Computational Biology, 11*(8), e1004273.

22. Francesca, G., Brambilla, M., Brutschy, A., Garattoni, L., Miletitch, R., Podevijn, G., et al. (2015). AutoMoDe-chocolate: Automatic design of control software for robot swarms. *Swarm Intelligence, 9*(2–3), 125–152.

23. Fujisawa, R., Dobata, S., Sugawara, K., & Matsuno, F. (2014). Designing pheromone communication in swarm robotics: Group foraging behavior mediated by chemical substance. *Swarm Intelligence, 8*(3), 227–246.

24. Garnier, S., Tâche, F., Combe, M., Grimal, A., Theraulaz, G. (2007). Alicein pheromone land: An experimental setup for the study of ant-likerobots. In: *Proceedings of the 2007 IEEE swarm intelligence symposium (SIS 2007)*, (pp. 37–44). Piscataway, NJ: IEEE.

25. Granovskiy, B., Latty, T., Duncan, M., Sumpter, D. J. T., & Beekman, M. (2012). How dancing honey bees keep track of changes: The role of inspector bees. *Behavioral Ecology, 23*(3), 588–596.

26. Gutiérrez, A., Campo, A., Monasterio-Huelin, F., Magdalena, L., & Dorigo, M. (2010). Collective decision-making based on social odometry. *Neural Computing and Applications, 19*(6), 807–823.

27. Hoff, N., Wood, R., & Nagpal, R. (2013). Distributed colony-level algorithm switching for robot swarm foraging. In A. Martinoli, F. Mondada, N. Correll, G. Mermoud, M. A. Hsieh, L. E. Parker, & K. Stoy (Eds.), *Distributed autonomous robotic systems, Springer tracts in advanced robotics* (Vol. 83, pp. 417–430). Berlin: Springer.

28. Jevtic, A., Gutiérrez, A., Andina, D., & Jamshidi, M. (2012). Distributed bees algorithm for task allocation in swarm of robots. *IEEE Systems Journal, 6*(2), 296–304.

29. Jones, C., & Matarić, M. J. (2003). Adaptive division of labor in large-scale minimalist multi-robot systems. In*Proceedings of the 2003 IEEE/RSJ international conference on intelligent robots and systems (IROS 2003)*(Vol. 2, pp. 1969–1974). Piscataway, NJ: IEEE Press.

30. Kazama, T., Sugawara, K., Watanabe, T. (2005). Traffic-like movement on a trail of interacting robots with virtual pheromone. In K. Murase, K. Sekiyama, T. Naniwa, N. Kubota, J. Sitte (Eds.), *Proceedings of the 3rd international symposium on autonomous minirobots for research and edutainment (AMiRE 2005)*, (pp. 383–388). Berlin: Springer.

31. Krieger, M. J. B., & Billeter, J. B. (2000). The call of duty: Self-organised task allocation in a population of up to twelve mobile robots. *Robotics and Autonomous Systems, 30*(1–2), 65–84.

32. Labella, T. H., Dorigo, & M., Deneubourg, J.-L. (2004). Efficiency and taskallocation in prey retrieval. In A. J. Ijspeert, M. Murata, & N. Wakamiya (Eds.), *Biologically inspired approaches to advanced information technology*, lecture notes in computer science (Vol. 3141, pp. 274–289). Berlin: Springer

33. Lee, J. H., Ahn, C. W. (2011). Improving energy efficiency in cooperative foraging swarm robots using behavioral model. In R. Abdullah, A. T. Khader, I. Venkat, L. P. Wong, K. G. Subramanian (Eds.), *Proceedings of the sixth international conference on bio-inspired computing: theories and applications (BIC-TA'11)*, (pp. 39–44). Piscataway, NJ: IEEE Press.

34. Lein, A., Vaughan, R. T. (2009). Adapting to non-uniform resource distributions in robotic swarm foraging through work-site relocation. In: *Proceedings of the 2009 IEEE/RSJ international conference on intelligent robots and systems (IROS2009)*, (pp. 601–606). Piscataway, NJ: IEEE Press.

35. Lemmens, N., de Jong, S., Tuyls, K., Nowe, A. (2008). Bee behaviour in multi-agent systems. In K. Tuyls, A. Nowe, Z. Guessoum, D. Kudenko (Eds.), *Adaptive agents and multi-agent systems III. Adaptation and multi-agent learning*, lecture notes in computer science, (Vol. 4865, pp. 145–156). Berlin: Springer.

36. Lerman, K., Jones, C., Galstyan, A., & Mataric, M. J. (2006). Analysis of dynamic task allocation in multi-robot systems. *The International Journal of Robotics Research*, *25*, 225–242.

37. Liu, W., & Winfield, A. F. (2010). Modelling and optimisation of adaptive foraging in swarm robotic systems. *The International Journal of Robotics Research*, *29*(14), 1743–1760.

38. Liu, W., Winfield, A. F., Sa, J., Chen, J., & Dou, L. (2007). Strategies for energy optimisation in a swarm of foraging robots. In E. Sahin, W. M. Spears, A. F. Winfield (Eds.), *Swarm robotics*, lecture notes in computer science, (Vol. 4433, pp. 14–26). Berlin: Springer.

39. Martinoli, A., Easton, K., & Agassounon, W. (2004). Modeling swarm robotic systems: A case study in collaborative distributed manipulation. *The International Journal of Robotics Research*, *23*(4), 415–436.

40. Mayet, R., Roberz, J., Schmickl, T., & Crailsheim, K. (2010). Antbots: A feasible visual emulation of pheromone trails for swarm robots. In M. Dorigo, M. Birattari, G. A. Di Caro, R. Doursat, A. P. Engelbrecht, D. Floreano, L. M. Gambardella, R. Groß, E. Sahin, H. Sayama, T. Stützle (Eds.), *Swarm intelligence*, lecture notes in computer science, (Vol. 6234, pp. 84–94). Berlin: Springer.

41. McFarland, D., & Spier, E. (1997). Basic cycles, utility and opportunism in self-sufficient robots. *Robotics and Autonomous Systems*, *20*, 179–190.

42. Miletitch, R., Trianni, V., Campo, A., & Dorigo, M. (2013). Information aggregation mechanisms in social odometry. In P. Liò, O. Miglino, G. Nicosia, S. Nolfi, M. Pavone (Eds.), *Proceedings of the twelfth European conference on the synthesis and simulation of living systems (ECAL 2013)*, (pp. 102–109). Cambridge, MA: MIT Press.

43. Miller, J. M., Wang, X. R., Lizier, J. T., Prokopenko, M., & Rossi, L. F. (2014). Measuring information dynamics in swarms. In M. Prokopenko

(Ed.), *Guided self-organisation: inception, emergence, complexity and computation* (Vol. 9, pp. 343–364). Berlin: Springer.

44. Nouyan, S., Groß, R., Bonani, M., Mondada, F., & Dorigo, M. (2009). Teamwork in self-organized robot colonies. *IEEE Transactions on Evolutionary Computation*, *13*(4), 695–711.

45. Pais, D., Caicedo-Núnez, C. H., & Leonard, N. E. (2012). Hopf bifurcations and limit cycles in evolutionary network dynamics. *SIAM Journal on Applied Dynamical Systems*, *11*(4), 1754–1784.

46. Parunak, H., & Brueckner, S. A. (2004). Engineering swarmings systems. In F. Bergenti, M. P. Gleizes, & F. Zambonelli (Eds.), *Methodologies and software engineering for agent systems, multiagent systems, artificial societies, and simulated organizations* (Vol. 11, pp. 341–376). Berlin: Springer.

47. Pérez-Uribe, A. (2001). Using a time-delay actor-critic neural architecture with dopamine-like reinforcement signal for learning in autonomous robots. In S. Wermter, J. Austin, D. Willshaw (Eds.), *Emergent neural computational architectures based on neuroscience*, lecture notes in computer science, (Vol. 2036, pp. 522–533). Berlin: Springer.

48. Pinciroli, C., Trianni, V., O'Grady, R., Pini, G., Brutschy, A., Brambilla, M., et al. (2012). ARGoS: A modular, parallel, multi-engine simulator for multi-robot systems. *Swarm Intelligence*, *6*(4), 271–295.

49. Pini, G., Brutschy, A., Pinciroli, C., Dorigo, M., & Birattari, M. (2013). Autonomous task partitioning in robot foraging: An approach based on cost estimation. *Adaptive Behavior*, *21*(2), 118–136.

50. Pitonakova, L., Crowder, R., & Bullock, S. (2014). Understanding the role of recruitment in collective robot foraging. In H. Lipson, H. Sayama, J. Rieffel, S. Risi, & R. Doursat (Eds.), *Proceedings of the fourteenth international conference on the synthesis and simulation of living systems (ALIFE 14)* (pp. 264–271). Cambridge, MA: MIT Press.

51. Ranjbar-Sahraei, B., Weiss, G., & Nakisaei, A. (2012). A multi-robot coverage approach based on stigmergic communication. In I. J. Timm & C. Guttmann (Eds.), *Multiagent system technologies*, lecture notes in computer science, (Vol. 7598, pp. 126–138). Berlin: Springer.

52. Reynolds, A. M., & Rhodes, C. J. (2009). The Lévy flight paradigm: Random search patterns and mechanisms.*Ecology*, *90*(4), 877–887.

53. Ribeiro, P. L., Helene, A. F., Xavier, G., Navas, C., & Ribeiro, F. L. (2009). Ants can learn to forage on one-way trails. *PloS one*, *4*(4), e5024.

54. Russell, R.A. (1999). Ant trails—An example for robots to follow? In: *Proceedings of the 1999 IEEE international conference on robotics and automation*, (Vol. 4, pp. 2698 – 2703). Piscataway, NJ: IEEE Press.

55. Sarker, M. O. F., & Dahl, T. S. (2011). Bio-Inspired communication for self-regulated multi-robot systems. In T. Yasuda (Ed.), *Multi-robot systems, trends and development* (pp. 367–392). InTech.

56. Schmickl, T., & Crailsheim, K. (2008). Throphallaxis within a robotic swarm: Bio-inspired communication among robots in a swarm. *Autonomous Robots*, *25*(1), 171–188.

57. Schmickl, T., & Hamann, H. (2010). BEECLUST: A swarm algorithm derived from honeybees. In Y. Xiao (Ed.),*Bio-Inspired Computing and Networking* (pp. 95–137). Abingdon: Routledge.

58. Schmickl, T., Möslinger, C., & Crailsheim, K. (2007). Collective perception in a robot swarm. In E. Sahin, W. M. Spears, & A. F. Winfield (Eds.), *Swarm robotics*, lecture notes in computer science, (Vol. 4433, pp. 144–157). Berlin: Springer.

59. Schmickl, T., Thenius, R., & Crailsheim, K. (2012). Swarm-intelligent foraging in honeybees: Benefits and costs of task-partitioning and environmental fluctuations. *Neural Computing and Applications*, *21*(2), 251–268.

60. Seeley, T. D. (1994). Honey bee foragers as sensory units of their colonies. *Behavioral Ecology and Sociobiology*, *34*(1), 51–62.

61. Seeley, T. D., & Morse, Ra. (1976). The nest of the honey bee (*Apis mellifera* L.). *Insectes Sociaux*, *23*, 495–512.

62. Seeley, T. D., Camazine, S., & Sneyd, J. (1991). Collective decision-making in honey bees: How colonles choose among nectar sources. *Behavioral Ecology and Sociobiology*, *28*, 277–290.

63. Seeley, T. D., Visscher, P. K., Schlegel, T., Hogan, P. M., Franks, N. R., & Marshall, Ja R. (2012). Stop signals provide cross inhibition in collective decision-making by honeybee swarms. *Science, 335*(6064), 108–111.

64. Serugendo, G. M., Gleizes, M. P., & Karageorgos, A. (2006). Self-organisation and emergence in MAS: An overview self-organisation. *Informatica, 30*, 45–54.

65. Shell, D. A., & Mataric, M. J. (2006). On foraging strategies for large-scale multi-robot systems. *Proceedings of the 2006 IEEE/RSJ international conference on intelligent robots and systems (IROS 2006)* (pp. 2717–2723). Piscataway, NJ: IEEE Press.

66. Sumpter, D. J. T., & Beekman, M. (2003). From nonlinearity to optimality: Pheromone trail foraging by ants.*Animal Behaviour, 66*(2), 273–280.

67. Tereshko, V., & Loengarov, A. (2005). Collective decision-making in honey bee foraging dynamics. *Computing and Information Systems Journal, 9*(3), 1–7.

68. Thenius, R., Schmickl, T., & Crailsheim, K. (2008). Optimisation of a honeybee-colony's energetics via social learning based on queuing delays. *Connection Science, 20*(2–3), 193–210.

69. Ulam, P., & Balch, T. (2004). Using optimal foraging models to evaluate learned robotic foraging behavior.*Adaptive Behavior, 12*(4), 213–222.

70. Valdastri, P., Corradi, P., Menciassi, A., Schmickl, T., Crailsheim, K., Seyfried, J., et al. (2006). Micromanipulation, communication and swarm intelligence issues in a swarm microrobotic platform. *Robotics and Autonomous Systems, 54*(10), 789–804.

71. Valentini, G., Hamann, H., & Dorigo, M. (2014). Self-organized collective decision making: The weighted voter model. In*Proceedings of the 13th international conference on autonomous agents and multiagent systems (AAMAS 2014)* (pp. 45–52). New York: ACM.

72. von Frisch, K. (1967). *The dance language and orientation of bees.* Cambridge: Harvard University Press.

73. Winfield, A. F. (2009). Towards an engineering science of robot foraging. In H. Asama, H. Kurokawa, & K. Sekiyama (Eds.), *Distributed Autonomous Robotic Systems 8* (pp. 185–192). Berlin: Springer.

74. Zahadat, P., Crailsheim, K., & Schmickl, T. (2013). Social inhibition manages division of labour in artificial swarm systems. In P. Liò, O. Miglino, G. Nicosia, S. Nolfi, & M. Pavone (Eds.), *Proceedings of the twelfth European conference on the synthesis and simulation of living systems (ECAL 2013)* (pp. 609–616). Cambridge, MA: MIT Press.

75. Zhang, D., Xie, G., Yu, J., & Wang, L. (2007). Adaptive task assignment for multiple mobile robots via swarm intelligence approach. *Robotics and Autonomous Systems*, *55*(7), 572–588.

Chapter 4

Travel Mode Detection Based on Neural Networks and Particle Swarm Optimization

Guangnian Xiao *, Zhicai Juan and Jingxin Gao

Antai College of Economics & Management, Shanghai Jiao Tong University, Shanghai 200052, China

ABSTRACT

The collection of massive Global Positioning System (GPS) data from travel surveys has increased exponentially worldwide since the 1990s. A number of methods, which range from rule-based to advanced classification approaches, have been applied to detect travel modes from GPS positioning data collected in travel surveys based on GPS-enabled smartphones or dedicated GPS devices. Among these approaches, neural networks (NNs) are widely adopted because they can extract subtle information from training data that cannot be directly obtained by human or other analysis techniques. However, traditional NNs, which are generally trained by back-propagation algorithms, are likely to be trapped in local optimum. Therefore, particle swarm optimization (PSO) is introduced to train the NNs. The resulting PSO-NNs are employed to distinguish among four travel modes (walk, bike, bus, and car) with GPS positioning data collected through a smartphone-based travel survey. As a result, 95.81% of samples are correctly flagged for the training set, while 94.44% are correctly identified for the test set. Results from this study indicate that smartphone-based travel surveys provide an opportunity to supplement traditional travel surveys.

KEYWORDS

global positioning system; neural networks; particle swarm optimization; travel mode

1. INTRODUCTION

With worsening traffic congestion, significant attention has been given to travel behavior research, as well as transportation demand analysis, to develop and assess transportation demand management strategies. In practice, travel surveys are generally utilized to collect the required information and data for travel demand modeling and analysis. However, conventional travel surveys, which are typically conducted using paper-and-pencil interviews, computer-assisted telephone interviews, or computer-assisted self-interviews, may significantly burden respondents because all relevant information, including travel modes, travel start and end times, trip destinations, and trip purposes, related to each trip should be reported throughout the entire survey period. Respondents of multi-day travel surveys experience severe fatigue, which could decrease the quality of retrieved data and adversely affect travel behavior analysis. Furthermore, some travel details are typically recalled or recorded approximately. Wolf *et al.* [1] revealed that respondents were significantly burdened with these surveys and indicated that trip timing and trip distance were prone to be rounded in traditional travel surveys. In addition, trip rates may be underestimated in conventional travel surveys. Du and Aultman-Hall [2] indicated that trips or trip chains with a short duration or distance were usually neglected.

It has been widely accepted that data collection efforts based on GPS technology can present evident advantages over traditional travel surveys. One of the most important benefits is the alleviation of the burden on respondents [1,3]. Burden reduction enables researchers to collect detailed travel information for a longer period without imposing additional burdens on respondents. Thus, the quality of the collected data is enhanced and researchers are allowed to investigate the dynamics of travel patterns over several days or weeks. For example, investigating the regularity and variability in trip destination decision, trip purpose frequencies, trip duration distributions, the number of daily trips, and temporal trip-making distributions over days of the week or even weeks of the month has become viable [4,5]. Additionally, employing GPS technology significantly improves the availability of accurate travel information. For example, trip rates are expected to be corrected through location acquisition with high precision [6,7]. GPS-enabled smartphones and dedicated GPS devices are typically utilized to record positioning data in GPS-based travel surveys. In recent years, travel surveys based on the latter have encountered more difficulties because of the following facts: (1) dedicated GPS devices are expensive; (2) the collected data are likely to be incomplete because respondents tend to forget to take the devices or charge them; (3) device distribution and recovery are required each time a new respondent is recruited; and (4) the number of respondents simultaneously participating in the survey is confined by the amount of available dedicated GPS devices [8].

The focus of the current study is to infer travel modes according to GPS positioning data. Rule-based algorithms [9,10,11] and advanced machine learning methods [12] are two frequently used methods to detect the travel mode. Although rule-based algorithms are readily understandable and applicable, these algorithms exhibit low generalizability [13]. That is to say, there exists a low extent to which the rules obtained from one case can be generalized to another, particularly when travel modes are detected in a city context with particular rules trained from data collected in another one. Zheng *et al.* [14] collected the GPS positioning data of 65 respondents for approximately 10 months and employed decision trees to correctly match 75.6% of trips. Subsequently, Rudloff and Ray [15] split 792 trajectories into training and test sets (70% to 30%) and applied a multilayer perceptron to properly infer travel modes for 82.70% of trips. Thereafter, Bolbol *et al.* [16] explored feature selection and employed a framework with a support vector machine (SVM) to achieve a promising accuracy of 88% for detecting travel modes. A more recent study by Broach *et al.* [17] used a multinomial logit (MNL) model to distinguish walk, bike, car and bus modes and correctly labelled 90.8% of trips. Table 1 shows the different methods applied to travel mode detection and summary of the corresponding accuracies.

Table 1. Summary of different methods for detecting travel modes.

Author/s	Method	Attributes	Sample size	Accuracy
Gong *et al.* [10]	Rule-based algorithm	GIS, speed, acceleration	340 segments	82.6%
Zheng *et al.* [14]	Decision trees	Speed, acceleration	65 respondents	75.6%
Rudloff and Ray [15]	Multilayer perceptron	GIS, speed, acceleration	792 trajectories	82.70%
Bolbol *et al.* [16]	Support vector machine	GIS, speed, acceleration	81 respondents	88%
Broach *et al.* [17]	Multinomial logit	GIS, speed, acceleration	926 segments	90.8%
Gonzalez *et al.* [18]	Neural networks	Speed, acceleration, data quality	114 trips	91.23%

A study by Gonzalez *et al.* [18] indicated that NNs are an appropriate approach for detecting travel modes from GPS positioning data since they can extract subtle information from training data that cannot be directly obtained by human or other analysis techniques. They used a number of statistics derived from the GPS data as input to the NNs. An individual looking at the same numbers may not necessarily be able to determine if the trip was by a certain mode. However, these features can be used to classify travel modes with NNs. In addition, the NNs have a powerful nonlinear pattern classification capacity, which makes NNs appropriate methods to extract travel modes from GPS data. They collected positioning data on 114 trips with smartphones and achieved a promising accuracy of 91.23%, which outperforms most of existing studies. However, the employed NNs were trained by back-propagation (BP) algorithms, which utilized gradient descent technique to train the parameters of the network. In this way, the fitness function is prone to be trapped into local optimum [19]. To address this issue, we include the particle swarm optimization (PSO) technique to search for a global optimum. Although PSO exhibits the partial optimism, which may cause lower accuracy at the regulation of its speed and the direction [20], the algorithm is a gradient-free approach, making the algorithm conceptually simple. During the iteration process of the search, only the globally best particle or the best particle in a group can transmit information to other particles, resulting in a high-speed search. In fact, a high searching speed is crucial to travel mode detection in GPS-based travel surveys, particularly when respondents are required to check the travel modes that are automatically derived from the GPS data. The resulting PSO-NNs are expected to achieve a favorable accuracy for travel mode detection.

The rest of this paper is organized as follows. Data collection and description are described in Section 2, and the theoretical and methodological background of NNs and PSO are presented in Section 3. The construction of the classifier is described in Section 4. Finally, the main conclusions and directions for future research are provided in Section 5.

2. DATA COLLECTION AND DESCRIPTION

2.1. Respondent Recruitment and Positioning App

The data used to detect travel modes were retrieved from a travel survey based on smartphones in Shanghai city from mid-October 2013 to mid-

February 2014. Respondents were typically limited to commuters in Shanghai to generate a relatively large number of trips and most of these respondents are recruited over the Internet. More details on the survey design are described in one of our previous studies [21]. A mobile app was developed to record GPS positioning data before the survey was conducted. Android and iOS were the platforms used in app development because of their considerable market penetration rates. The app records UTC time, latitude, longitude, altitude, the instantaneous speed, bearing, the number of satellites in view and Horizontal Dilution of Precision (HDOP) once every second. HDOP represents how these satellites in view are arranged in the sky. HDOP is a factor in determining the relative accuracy of a horizontal position. The smaller the HDOP number is, the better the geometric arrangement is. In practice, battery consumption might be an intractable problem with such a high recording frequency, although this frequency increases the accuracy of travel mode imputation. Thus, we gave each respondent a mobile power supply to avoid battery drainage. The package is expected to motivate more respondents to participate in the travel survey. In addition, the app automatically shut down when the smartphone remains stationary for over five minutes and automatically restarts when the smartphone moves again. This function is designed to decrease battery consumption to a minimum extent without adversely effecting on the regular location recording during trips.

2.2. Requirements of the Travel Survey

Respondents were asked to start the app to enable automatic location acquisition before their first trip, and keep the device running until their final arrival back home each day during the entire survey period. Respondents were requested to shut down the app at home to minimize battery consumption as much as possible and relieve privacy issues on the respondents, particularly those who are unwilling to be positioned at home. Respondents were asked to upload the positioning data to a database on our server with a single touch and recommended to upload the data in a Wi-Fi environment to decrease transmission cost.

Single-day GPS positioning data were first divided into trips, where each was split into single-mode segments (known as segments) prior to detecting travel modes. The approach utilized for trip division is described in one of our previous studies [22]. The basic assumption in the study is that trip ends are characterized by either GPS point clustering or sudden direction change in the context of normal GPS location acquisition, while they display a dwell time exceeding the threshold during GPS signal loss. The segments are extracted from trips based on the approach suggested by Gong *et al.* [10]. After automatically deriving the segments from GPS trajectories, we contacted respondents by telephone and asked them to check these segments. The partition of the segments was corrected when necessary. The respondents were also asked to check the travel modes of each segment. Such a travel survey is also called prompted recall survey. As expected the respondents do not experience a significant burden because most trip ends and travel modes can be inferred correctly. The travel mode of each segment that was validated by the respondents is taken as the "ground truth" in the subsequent study.

2.3. Sample Data

Each respondent was requested to participate in the travel survey for at least five continuous days. A total of 113 respondents participated in this survey but only 102 completed it. Those who did not complete the survey quit because of various reasons such as smartphone failure and privacy issues. The number of respondents who completed this survey was categorized based on the GPS data uploaded within a given number of days: 83 respondents with five days and 19 respondents with six to eight days. In addition, 73 of the 102 respondents kept their survey days continuous. Although this continuity requirement does not substantially affect the travel mode detection, this can prompt the respondents to complete the survey quickly. A total of 535 days of GPS data were collected.

In the current study, we distinguished among walk, bike, bus and car modes. Signal loss during subway trips is a serious issue. Thus, it is not included in the analysis. The signal loss for the other four travel modes is handled as follows. The GPS positioning data were expected to be retrieved with a regular interval of one second. Thus, the duration of each missing trajectory can be calculated easily. If the duration of any missing trajectory is less than 10 s, it is compensated with a straight line by itself with lost positioning points located uniformly on this line. Based on the experiences gained in the travel survey, 10 s is an appropriate criteria that could balance respondent burdens with the accuracy required by the travel survey. The positioning data that incorporated missing trajectories greater than 10 s were not included in the current analysis to achieve high data accuracy. A total of 1654 segments were extracted from the recorded trajectories, and all of these segments were validated by respondents. The number of segments for the walk, bike, car and bus modes is 820, 148, 326, and 360, respectively. The walk mode exhibits a maximum percentage with an overwhelming advantage because walk segments usually play a transitional role when intermodal transfers are made. For example, a person usually needs to walk to a subway station after getting off a bus when the bus and subway are sequentially taken during a trip. In addition, walk segments usually exist before the bus segment and after the subway segment. Based on the preceding analysis, this trip is generally broken down into three walk segments, one subway segment and one bus segment. Therefore, effectively imputing walk segments is crucial in the overall detection performance.

3. TRAVEL MODE DETECTION

3.1. Feature Selection

A feature set that incorporates a high distinctiveness provides an opportunity to greatly improve the performance of classification. According to existing studies, the average speed, median speed, average absolute acceleration, travel distance and the 95th percentile speed are typically used to infer travel modes (as shown in Figure 1a–e). However, distinguishing bus segments from car segments is difficult when only the aforementioned speed-related features are utilized [9,18]. In most cases, the inclusion of a transit network layer can significantly improve the distinction degree of bus segments from car segments [23]. However, the transit network layer is not available for most researchers, and we also encountered the same difficulty. In addition, the bus network is updated every month and even every day at times in a megacity such as Shanghai. A timely update is required to effectively infer bus segments. Thus, it is costly to maintain an up-to-date GIS layer of the bus network. All these factors motivated us to develop a new feature called "low-speed point rate", *i.e.*, the rate of points with a speed of less than 1 m/s, which is expected to capture the characteristics of the periodical stops of buses. In fact, we investigated four types of critical speeds, *i.e.*, 0.5 m/s, 1.0 m/s, 1.5 m/s and 2.0 m/s, for this feature. According to the two-sample Kolmogorov-Smirnov test for each "low-speed rate" (0.5 m/s, 1 m/s, 1.5 m/s and 2 m/s) between bus segments and car segments, 1 m/s achieves the minimum *p* value ($p = 1.9456 \times 10^{-7}$). The p values of 0.5 m/s, 1.5 m/s and 2 m/s are 8.9686×10^{-7}, 9.7462×10^{-5}, and 7.1505×10^{-4}, respectively. Thus, 1 m/s is applied because this value can best differentiate these two segments (as shown in Figure 1f). The distribution of the six features is shown in Figure 1.

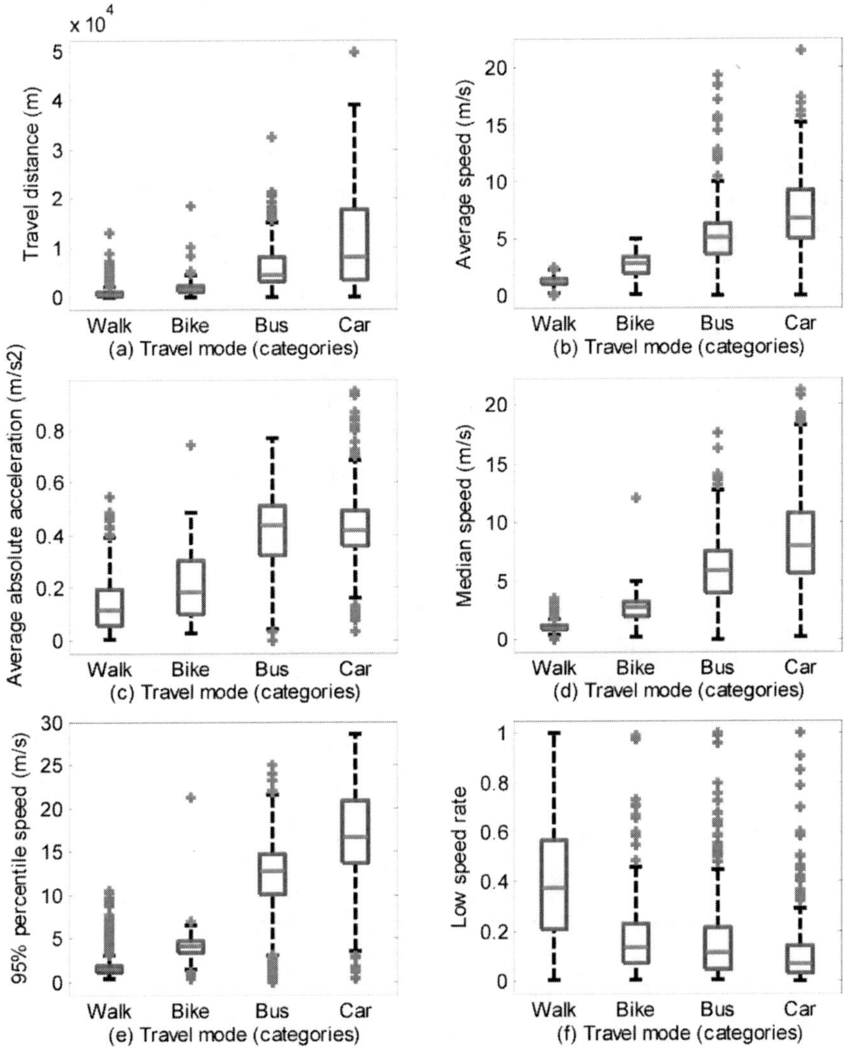

Figure 1. Distribution of six features.

3.2. Methodology

3.2.1. Neural Networks

A neural network consists of a selection of input, hidden and output neurons. Generally, any two neurons between the adjacent layers are directly correlated. The neural network collects the inputs, extracts any useful information through hidden neurons and produces the output neurons that can be used to classify the input sample. In the current study, the weights are trained by the PSO algorithm, which is described in the next subsection. The trained neural network, which is represented by a series of weighting matrices, is employed to evaluate the data streams with correct outputs that the researchers are unaware of. The neural network can be highly flexible because various types of data are readily fitted when the multiple settings of a neural work are adjusted, including the amount of hidden layers, number of neurons in each layer, and learning rate. In this study, we employed a three-layer neural network to detect travel modes from the GPS positioning data (as shown in Figure 2). Additionally, we applied a commonly used logistic function as the activation function.

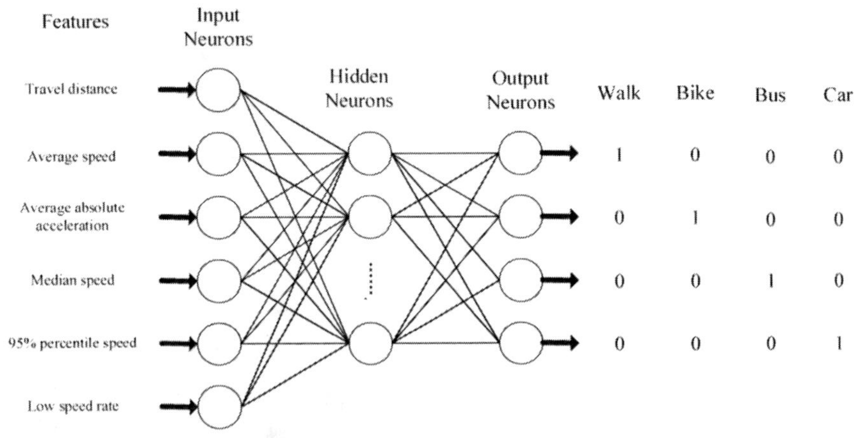

Figure 2. Three-layer neural network.

3.2.2. Particle Swarm Optimization

PSO is essentially an evolutionary computation technique proposed by Kennedy and Eberhart [24]. PSO originates from a natural system and incorporates a global search ability. The algorithm is developed according to studies based on the social behavior of animals, such as fish schooling and bird flocking. To improve the probability of convergence to a global optimal, we employed a common ring topology [25]. In this topology, each particle corresponds to a specific group of particles, which consists of the particle in question and its immediate adjacent neighbors. In addition, the connections within a group are undirected and unweighted. The algorithm is initialized with a swarm of random solutions. Each initialized solution, also known as a particle, is represented by a random initial location and velocity. At each step *n*, the velocity is updated according to the following equation:

$$\mathbf{v}_i(n+1) = \omega \times \mathbf{v}_i(n) + c_1 \times rand_1 \times (\mathbf{p}_{pbest\_i} - \mathbf{x}_i(n)) + c_2 \times rand_2 \times (\mathbf{p}_{lbest\_i} - \mathbf{x}_i(n)) \quad (1)$$

In the velocity updating equation, the new velocity vector vi(n+1) corresponding to particle *i* can be obtained by computing the sum of three components. The first component is known as "inertia" and the parameter ω is the inertia weight. The second component is known as "individual knowledge", which represents the effect of the known best position vector Ppbest_i that particle *i* has determined on the current velocity. The third component is known as "group knowledge", which indicates the effect of the known best position Plbest that all the particles in the group have found on the current velocity. The acceleration parameters c1 and c2 are included in the second and third components to adjust these two types of impact, whereas two random numbers rand1 and rand2, which have a uniform distribution between 0 and 1, are applied to provide random search.

After the new velocity is calculated, the particle position is updated based on the following equation:

$$\mathbf{x}_i(n+1) = \mathbf{x}_i(n) + \mathbf{v}_i(n+1) \quad (2)$$

where xi(n) indicates the position vector of particle *i* in the previous iteration.

For each iteration, we need to evaluate the fitness of each particle based on the desired optimization purpose. To evaluate a neural network, the root-mean-square-error (RMSE) is utilized to describe the average detection error and can be calculated according to the following equation:

$$RMSE = \sqrt{\frac{1}{nTr}\sum_{i=1}^{nTr}\sum_{j=1}^{nOut}(t_{i,j} - y_{i,j})^2} \quad (3)$$

where *nTr* represents the amount of the training samples, *nOut* is the amount of network outputs, and t_{ij} and y_{ij} are the *j*th derived output and reported output corresponding to the *i*th training sample, respectively. For example, if the reported travel mode of the *i*th training sample is walk, y_{i1} is equal to 1. Accordingly, if the derived travel mode of the *i*th training sample is bus, t_{i1} and t_{i3} are equal to 0 and 1, respectively. The fitness function is defined as follows:

$$fitness = \frac{1}{1 + RMSE} \quad (4)$$

Fitness can be used to measure the performance of a neural network. According to the evaluation results of the initial swarm, the personal best of each particle and the local best of the group will be stored. Obviously, the personal best of each particle is equal to its initial position in the initial state, whereas the local best records the position of the particle that achieves the best fitness value calculated with Equation (4) in a group. The optimization process continues by updating the personal best and the local best with Equations (1) and (2). If the fitness of a particle is greater than its personal best, the local best is replaced with the particle. Similarly, if the fitness of a particle is greater than the local best, the local best is replaced with the current particle. The process does not stop until the criterion set by the user is met. The criterion may include the threshold of the RMSE or the maximum iteration number of the optimization process.

4. DETECTING TRAVEL MODES WITH PSO-NNS

According to the sample partition applied by Feng and Timmermans [13], three quarters of the total sample is taken as the training set, whereas the remaining one quarter of the total sample is regarded as the test set. In this way, the number of samples in training set and test set is 1240 and 414, respectively. In the current study, the acceleration parameters c1 and c2are taken to be constants 1 and 2 according to Lin and Hsieh [19]. To increase the global search ability at the beginning of the iteration process and improve the local search ability near the end of the iteration process, we used a linearly decreasing inertia weight, which is represented as follows:

$$\omega = (\omega_s - \omega_e)(t_{\max} - t) / t_{\max} + \omega_e \tag{5}$$

where t_{max} and t are the maximum number of iterations and the current iteration times, and ω_s and ω_e are the initial and final values of the inertia weight, respectively. Following the recommendation of Shi and Eberhart [26], we apply the values of 0.9 and 0.4 for ω_s and ω_e. In addition, the maximum number of iterations t_{max} = 5000 is used in this study [27].

We defined the population size as 50 particles. A particle in the current study is encoded for a vector. For a neural network involved, each particle represents all weights of the structure of the neural network. If the number of neurons in the hidden layer in Figure 2 is 5, the neural network has the structure of 7-6-4, including a bias unit in both of the first two layers. In this case, each particle is a vector with a length of 66. Before implementing the classification efforts, we rescaled the four dimensions (travel distance, average speed, median speed and 95% percentile speed) to the range [0, 1] so that all the features have the same range. The formula is given as follows:

$$x_i^{'} = \frac{x_i - \min(\mathbf{x})}{\max(\mathbf{x}) - \min(\mathbf{x})} \tag{6}$$

where x'i and xi are the *i*th original and normalized sample values for a specific feature, respectively, and **x** is the vector consisting of all sample values for the feature. To avoid a particle exceeding the range of the problem, x_{max} = 1 and x_{min} = 0 are used for each dimension. Accordingly, the maximum velocity v_{max} = 1 and v_{min} = −1 are used in this study [28]. Figure 3illustrates the pseudocode of a PSO using a ring topology. In this study, a "wrap-around" ring topology is used. In other words, the last particle is the neighbor of the first particle and *vice versa*. The function neighborhood(.) returns the best personal best in the neighborhood of particle *i*.

As stated above, we employed a three-layer neural network to detect travel modes. We tested multiple NNs that included the number of hidden neurons ranging from 1 to 20 [27]. For each neural network, the classification accuracy for the test set is shown in Figure 4. The classification accuracy is represented as the ratio of the number of samples that were correctly classified to the total number of samples. The highest accuracy of 94.44% was achieved for the neural network that incorporates 12 hidden neurons. Figure 4 indicates that the classification accuracy increases rapidly when the number of neurons increases from 1 to 8, whereas it gradually decreases after the number of neurons exceeds 12. When 12 neurons are applied, the classification accuracy reaches 94.44%, which is higher than the results of most existing studies [29]. An exception is the study by Stopher *et al.* [9], which achieves an accuracy of approximately 95%. However, it is difficult to compare different studies because the accuracy of travel mode detection depends on the spatial context in which GPS is used, the number of identified transportation modes, type of input variables and data used for validation [13]. Therefore, we applied the neuron network that includes 12 neurons in the hidden layer.

```
Randomly initialize population of particles, and derive p_pbest_i
and p_lbest_i for each particle
for k = 1 to the maximum number of iterations do
    update ω according to equation (5)
    for i = 1 to Population Size do
    if fitness(x_i) > fitness(p_pbest_i) then p_pbest_i = x_i
    end
    for i = 1 to Population Size do
    P_lbest_i =neighborhoodBest(p_pbest_i-1, p_pbest_i, p_pbest_i+1,)
    end
    for i = 1 to Population Size do
    update particle position according to equation (1) and (2)
    end
end
```

Figure 3. Pseudocode of a PSO using a ring topology.

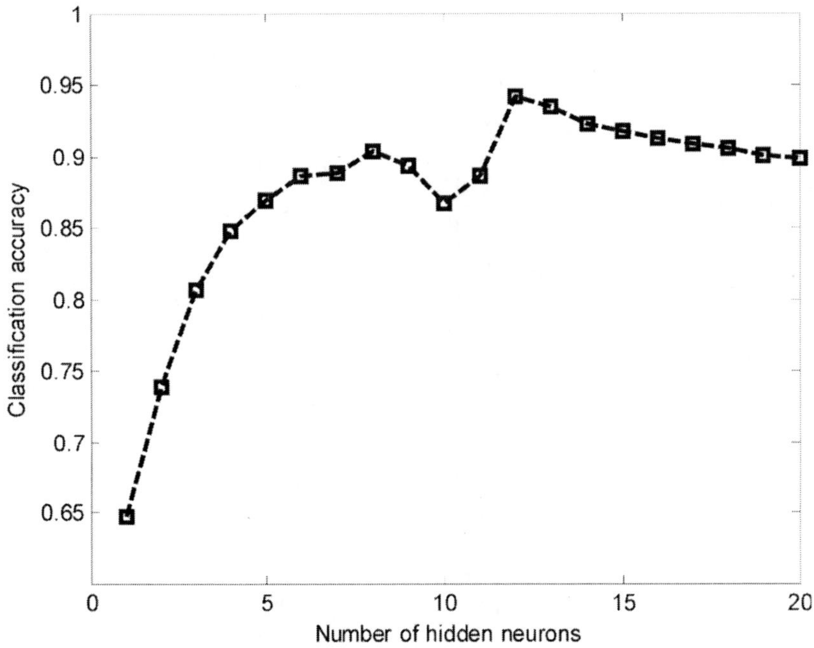

Figure 4. Classification accuracy for different number of hidden neurons.

To further explore the performance of the classifier, we calculated the confusion matrix. Recall and precision are used to measure the classification capacity of the classifier and their definitions are described by Forman [30] in detail. As shown inTable 2, the highest recall is achieved for segments with walk modes and it reaches 96.59% for the test set and 97.24% for the training set. This result is predictable because several features, such as the average speed and travel distance, exhibit an obviously distinct distribution between walk modes and other modes. Recall for segments with bus and car modes is relatively low among the four modes. In addition, most bus segments that are incorrectly flagged are identified as car modes, and *vice versa*. This result was also obtained by other studies [9]. Regarding precision, the lowest value is obtained for bike segments, which may be due to the fact that the speed related features for bike modes are between walk modes and bus/car modes.

Table 2. Confusion matrix of the neural network.

		Training Set: Detected					Test Set: Detected				
		Walk	Bike	Bus	Car	Recall (%)	Walk	Bike	Bus	Car	Recall (%)
Reported	Walk	598	10	4	3	97.24	198	2	3	2	96.59
	Bike	2	107	1	1	96.40	1	35	1	0	94.59
	Bus	3	3	230	8	94.26	1	2	76	3	92.68
	Car	2	5	10	253	93.70	1	1	6	82	91.11
Precision (%)		98.84	85.60	94.26	95.47	95.81	98.51	87.50	88.37	94.25	94.44

In comparing PSO-NNs with other frequently used machine learning methods, we chose several representative classifiers, consisting of SVM, MNL, and neural networks with back-propagation (BP-NNs), to detect the travel modes. The SVM classifier is implemented with package "probsvm" in R, with one-*versus*-rest used in the multiclass method. The MNL model [31] is used with its primary version and all the features are alternative-specific attributes. BP-NNs are performed with the same acceleration parameters, inertia parameter strategy, population size, velocity limits, and maximum number of iterations and layers of neurons as the PSO-NNs. In addition, BP-NNs are also tested with 1 to 20 neurons in the hidden layer, and the highest accuracy with 15 neurons was achieved. All these approaches share the same training data and test data, ensuring that the classification capacity is comparable among classifiers.

The results of the comparison of these classifiers are shown in Table 3 and they are predictable to a certain extent. For example, although the MNL model is readily understandable and practicable, this model does not fit the problem well because the basic assumption of independence of irrelevant alternatives (IIA) may not hold for this issue. The comparison between BP-NNs and PSO-NNs indicates that implementing PSO algorithm evidently improves the classification ability of the neural network. More specifically, the classification accuracy for the training set is improved from 91.85% to 95.81%, while that for test set is improved from 89.37% to 94.44%. According to McNemar's test, the resulting M value (M = 9.30) is greater than the critical value 3.84 (95% confidence level). Thus, this test rejects the null hypothesis that BP-NNs and PSO-NNs have the same classification errors. This result indicates that it is more reliable to apply the classifier of PSO-NNs to evaluate new samples.

5. SUMMARIES AND CONCLUSIONS

The current study employed PSO-NNs to distinguish the four travel modes (walk, bike, bus, and car) with GPS positioning data collected through a smartphone-based travel survey. The derived travel modes for each segment were compared with the "ground truth", which was obtained by respondents validating or correcting the primary travel information that was automatically detected from trajectories uploaded by the respondents themselves. Based on comparison of PSO-NNs with several representative classifiers, the PSO-NNs achieved the best classification accuracy of 94.44% for the test set and 95.81% for the training set.

Table 3. Comparison among several classifiers for detecting travel modes.

	Training Set		Test Set	
	# Segments Correctly Flagged	Accuracy (%)	# Segments Correctly Flagged	Accuracy (%)
SVM	1107	89.27	356	85.99
MNL	969	78.15	299	72.22
BP-NNs	1142	91.85	370	89.37
PSO-NNs	1188	95.81	391	94.44

By improving the classification accuracy of travel modes with PSO-NNs, respondent burdens are expected to be further reduced and the survey duration can be prolonged without requiring additional effort from the respondents. With the improvement of the classification accuracy for travel mode detection and the decrease of burdens on respondents, the smartphone-based travel surveys provide an opportunity to supplement traditional travel surveys. According to the investigation of the confusion matrix, newly targeted features may be included to decrease the classification errors between bus and car modes in future studies.

ACKNOWLEDGMENTS

This research has been supported by the National Natural Science Foundation of China (51478266). The authors would like to express their appreciation to anyone who has provided suggestions and comments on this paper.

AUTHOR CONTRIBUTIONS

Guangnian Xiao and Zhicai Juan conceived and designed the data collection methods; Jingxin Gao performed the data collection process; Guangnian Xiao and Jingxin Gao wrote the code and analyzed the data; Zhicai Juan contributed data analysis tools; Guangnian Xiao wrote the paper. All authors have read and approved the final manuscript.

CONFLICTS OF INTEREST

The authors declare no conflict of interest.

REFERENCES

1. Wolf, J.; Hallmark, S.; Oliveira, M.; Guensler, R.; Sarasua, W. Accuracy issues with route choice data collection by using global positioning system. *Transp. Res. Rec. J. Transp. Res. Board* **1999**, *1660*, 66–74.

2. Du, J.; Aultman-Hall, L. Increasing the accuracy of trip rate information from passive multi-day GPS travel datasets: Automatic trip end identification issues. *Transp. Res. Part A Policy Pract.* **2007**, *41*, 220–232.

3. Zhou, J.; Golledge, R. Real-time tracking of activity scheduling/schedule execution within a unified data collection framework. *Transp. Res. Part A Policy Pract.* **2007**, *41*, 444–463.

4. Nijland, L.; Arentze, T.; Timmermans, H. Multi-day activity scheduling reactions to planned activities and future events in a dynamic model of activity-travel behavior. *J. Geogr. Syst.* **2014**, *16*, 71–87.

5. Arentze, T.A.; Ettema, D.; Timmermans, H.J. Location choice in the context of multi-day activity-travel patterns: Model development and empirical results. *Transp. A Transp. Sci.* **2013**, *9*, 107–123.

6. Forrest, T.L.; Pearson, D.F. Comparison of trip determination methods in household travel surveys enhanced by a global positioning system. *Transp. Res. Rec. J. Transp. Res. Board* **2005**, *1917*, 63–71.

7. Stopher, P.; FitzGerald, C.; Xu, M. Assessing the accuracy of the sydney household travel survey with GPS.*Transportation* **2007**, *34*, 723–741.

8. Maruyama, T.; Mizokami, S.; Hato, E. A Smartphone-Based Travel Survey Trial Conducted in Kumamoto, Japan: An Examination of Voluntary Participants' Attributes. In Proceedings of the Transportation Research Board 93rd Annual Meeting, Washington, DC, USA, 12–16 January 2014.

9. Stopher, P.R.; Jiang, Q.; FitzGerald, C. Processing GPS Data from Travel Surveys. In Proceedings of the 2nd International Colloqium on the Behavioural Foundations of Integrated Land-Use and Transportation Models: Frameworks, Models and Applications Toronto, Toronto, ON, Canada, 12–15 June 2005.

10. Gong, H.; Chen, C.; Bialostozky, E.; Lawson, C.T. A GPS/GIS method for travel mode detection in New York city.*Comput. Environ. Urban Syst.* **2012**, *36*, 131–139.

11. Chen, C.; Gong, H.; Lawson, C.; Bialostozky, E. Evaluating the feasibility of a passive travel survey collection in a complex urban environment:

Lessons learned from the New York city case study. *Transp. Res. Part A Policy Pract.* **2010**,*44*, 830–840.

12. Kusakabe, T.; Asakura, Y. Behavioural data mining of transit smart card data: A data fusion approach. *Transp. Res. Part C Emerg. Technol.* **2014**, *46*, 179–191.

13. Feng, T.; Timmermans, H.J. Comparison of Advanced GPS Data Imputation Algorithms for Detection of Transportation Mode and Activity Episode. In Proceedings of the Transportation Research Board 93rd Annual Meeting, Washington, DC, USA, 12–16 January 2014.

14. Zheng, Y.; Chen, Y.; Li, Q.; Xie, X.; Ma, W.-Y. Understanding transportation modes based on GPS data for web applications. *ACM Trans. Web (TWEB)* **2010**, *4*, 1–36.

15. Rudloff, C.; Ray, M. Detecting Travel Modes and Profiling Commuter Habits Solely Based on GPS Data. In Proceedings of the Transportation Research Board 89th Annual Meeting, Washington, DC, USA, 10–14 January 2010.

16. Bolbol, A.; Cheng, T.; Tsapakis, I.; Haworth, J. Inferring hybrid transportation modes from sparse GPS data using a moving window svm classification. *Comput. Environ. Urban Syst.* **2012**, *36*, 526–537.

17. Broach, J.; McNeil, N.W.; Dill, J. Travel Mode Imputation Using GPS and Accelerator Data from Multi-Day Travel Survey. In Proceedings of the Transportation Research Board 93rd Annual Meeting, Washington, DC, USA, 12–16 January 2014.

18. Gonzalez, P.A.; Weinstein, J.S.; Barbeau, S.J.; Labrador, M.A.; Winters, P.L.; Georggi, N.L.; Perez, R. Automating mode detection for travel behaviour analysis by using global positioning systems-enabled mobile phones and neural networks. *IET Intell. Transp. Syst.* **2010**, *4*, 37–49.

19. Lin, C.-J.; Hsieh, M.-H. Classification of mental task from eeg data using neural networks based on particle swarm optimization. *Neurocomputing* **2009**, *72*, 1121–1130.

20. Selvi, V.; Umarani, D.R. Comparative analysis of ant colony and particle swarm optimization techniques. *Int. J. Comput. Appl. (0975–8887)* **2010**, *5*.

21. Xianyu, J.; Juan, Z.; Xiao, G.; Fu, X. Smartphone-based travel survey: A pilot study in China. *Mob. Technol. Act.-Travel Data Collect. Anal.* **2014**, *209*.

22. Xiao, G.; Juan, Z.; Gao, J. Inferring Trip Ends from GPS Data Based on Smartphones in Shanghai. In Proceedings of the Transportation Research Board 94th Annual Meeting, Washington, DC, USA, 11–15 January 2015.

23. Joseph, B.; Nathan, W.M.; Jennifer, D. Travel Mode Imputation Using GPS And Accelerator Data From Multi-Day Travel Survey. In Proceedings of the Transportation Research Board 93rd Annual Meeting, Washington, DC, USA, 12–16 January 2014.

24. Kennedy, J.; Eberhart, R. Particle Swarm Optimization. In Proceedings of IEEE International Conference on Neural Networks, Perth, Australia, 27 November–1 December 1995; pp. 1942–1948.

25. Kennedy, J.; Kennedy, J.F.; Eberhart, R.C.; Shi, Y. *Swarm Intelligence*; Morgan Kaufmann: Burlington, MA, USA, 2001.

26. Shi, Y.; Eberhart, R.C. Empirical Study of Particle Swarm Optimization. In Proceedings of the 1999 Congress on Evolutionary Computation, CEC 99, Washington, DC, USA, 6–9 July 1999.

27. Cai, X.; Zhang, N.; Venayagamoorthy, G.K.; Wunsch, D.C. Time series prediction with recurrent neural networks trained by a hybrid pso-ea algorithm. *Neurocomputing* **2007**, *70*, 2342–2353.

28. Schutte, J.F.; Groenwold, A.A. A study of global optimization using particle swarms. *J. Glob. Optim.* **2005**, *31*, 93–108.

29. Shen, L.; Stopher, P.R. Review of GPS travel survey and GPS data-processing methods. *Transp. Rev.* **2014**, *34*, 316–334.

30. Forman, G. An extensive empirical study of feature selection metrics for text classification. *J. Mach. Learn. Res.* **2003**, *3*, 1289–1305.

31. McFadden, D. Conditional logit analysis of qualitative choice behavior. *Front. Econom.* **1973**, 105–142.

Chapter 5

Artificial Searching Swarm Algorithm and Its Performance Analysis

Tanggong Chen, Wang Guo, Zhijian Gao

Province-Ministry Joint Key Laboratory of Electromagnetic Field and Electrical Apparatus Reliability, Hebei University of Technology, Tianjin, China

ABSTRACT

Artificial Searching Swarm Algorithm (ASSA) is a new optimization algorithm. ASSA simulates the soldiers to search an enemy's important goal, and transforms the process of solving optimization problem into the process of searching optimal goal by searching swarm with set rules. This work selects complicated and highn dimension functions to deeply analyse the performance for unconstrained and constrained optimization problems and the results produced by ASSA, Genetic Algorithm (GA), Particle Swarm Optimization (PSO), Artificial Fish-Swarm Algorithm (AFSA) have been compared. The main factors which influence the performance of ASSA are also discussed. The results demonstrate the effectiveness of the proposed ASSA optimization algorithm.

KEYWORDS

Artificial Searching Swarm Algorithm; Bionic Intelligent Optimization Algorithm; Optimization; Evolutionary Computation; Swarm Intelligence

1. INTRODUCTION

Optimal design is always the goal in the engineering design fields. The traditional optimization algorithms often depend on the quality of the objective function, but many objective functions are usually highly non-linear, steep, multi-peak, non-differential or even discontinuous, and have many continual or discrete parameters. Almost all the problems need vast amount of computation. The traditional optimization techniques are incapable to solve these problems.

Most animals and insects show the amazing abilities of completing complex behaviors. Since 1940s, the optimization design problems in the engineering fields have been solved by using the inspiration of the biological systems, and meantime a new type of optimization method — Bionic Intelligent Optimization Algorithm (BIOA) is found. At present, the popular bionic intelligent optimization algorithms are Genetic Algorithm (GA) [1], Ant Colony Algorithm (ACA) [2], Particle Swarm Optimization (PSO) [3], Artificial Fish-Swarm Algorithm (AFSA) [4], and Shuffled Frog Leaping Algorithm (SFLA) [5]. These bionic algorithms have become a research focus in the fields of intelligent optimization.

Simulating a group soldiers to complete the task for searching the enemy's important goals, a new bionic intelligent optimization algorithm—Artificial Searching Swarm Algorithm (ASSA) is presented and simply discussed [6]. In ASSA, the process of solving the optimization problem is transformed into the process of finding the optimal solution for searching swarm with set rules, then through the search and cooperation of searching individuals the optimal solution is expected to find. This work selects complicated and highn dimension functions to deeply analyse the performance for optimizing multivariable functions and the results produced by ASSA, GA, PSO, AFSA have been compared, and the main factors which influence the performance of ASSA are also discussed. Simulation tests showed that ASSA is an efficient optimization algorithm.

2. ABSTRACT MODEL OF BIOA

Although the characters are different, BIOA manifests the great similarity in the structure, operation model, research method and the research content. It provides the possibility for the establishment of abstract model.

The abstract model of BIOA can be described as follows: the swarm is formed by the individuals, relies on the specific evolution rules to generate the reborn swarm (such as GA) or change the individual position (such as PSO, AFSA), and goes with the iteration of the algorithm, the current solution evolves unceasingly along with the change of the swarm and then approaches the optimal solution gradually [7,8].

The process of the algorithms is shown as follows:

1) Set the parameters, initialize the swarm randomly and calculate the fitness values;

2) According to the set rules, renew swarm or its position, generate a group of solutions, and calculate the fitness of individuals;

3) Through comparison, obtain the optimal fitness value and take note;

4) Judge whether or not the terminal condition is satisfied. If satisfy, end the iteration; otherwise, return to 2.

In this abstract model, the rules of renewing swarm decide the algorithm performance. These set rules restrict the individual behaviors, and form the unique characteristics of the algorithm to be different from the other algorithms.

In BIOA the bulletin board is generally used to record the optimal individual's historical states. Through the iteration of the algorithm, each individual compares its own condition with bulletin board's condition, and replaces the information when its value is better. Thus the bulletin board can record the historical best solution all along. After algorithm iteration finished, the optimal solution and the related information can be obtained from the bulletin board.

3. ARTIFICIAL SEARCHING SWARM ALGORITHM

3.1. Basic Idea of ASSA

According to the abstract model of BIOA, ASSA uses the swarm composed of searching individuals as the executive to implement the searching task, and transforms the process of solving optimal design problem into the process of implementing relevant task and searching optimal goal of the searching swarm with set rules. Through the uninterrupted search of searching swarm, the better solution evolves unceasingly and approaches the optimal solution gradually, so the corresponding optimization problem is solved.

According to the above ideas, how to set the rules becomes the key problem. The rules of the algorithm are the limitation of the searching individual's behavior, which decide directly the searching swarm's movement mechanism, and affect the efficiency and the success of the algorithm.

The searching individual behaviors can be described as follows:

1) Synergetic move rule: basic communication relation is kept between searching individuals. If an individual receives a call sent by the other individuals, it moves forward to the called individual's position by a step with a certain probability;

2) Reconnaissance move rule: If the individual does not receive the call, it implements reconnaissance according to it's and swarm's historical experience. If finds a better goal, then moves forward to the position by a step;

3) Stochastic move rule: If the searching individual does not receive a call, and does not find a better goal, it moves a step randomly.

If finds a better goal during above moves, sends a call to searching swarm.

Three rules of the ASSA have their own characteristics. Rule 1 can compel the searching swarm moving forward to the better goal, and accelerate the convergence speed of the algorithm; Rule 2 plays a main role in finding the better goals, pulls the algorithm away from the local solution, and has the global adjusting function; Rule 3 plays a supporting role in moving swarm. The above analyses have been confirmed by simulation tests.

The running flow of ASSA is shown as follows:

1) Set the parameters, initialize the swarm randomly and calculate the fitness value;

2) Iteration counter add 1, deal with the individuals in turn as follows:

    a) If receive a call, then move forward to the called individual by a step with a certain probability;

    b) Otherwise, according to its own and swarm's historical experience, implement the reconnaissance. If find a better goal, move forward to the better goal by a step;

    c) Or move by a step randomly.

If find the better goal during above moves, send a call to searching swarm.

3) Calculate the fitness value. Compare with the best fitness of swarm and each individual respectively, if better, log on the bulletin board;

4) Determine whether or not to satisfy the conditions of termination, if satisfy, then end the iteration; otherwise, return to 2.

3.2. Pseudo-Code Description of the ASSA

For explaining the behavior rule clearly, by using objectoriented technology, a searching individual can be described as a C++ class as follows:

class Artificial_Searching_Individuals

{

float X[n]; //individual's position.

float AS_step; //the distance moved for a step float Pc; //synergetic parameter, the value of probability to implement the synergetic move rule float AS_fitness(); //the object function void AS_synergetic(); //the behavior of synergetic void AS_reconnaissance(); //the behavior of

```
reconnaissance void AS_move(); //the behavior of move
Artificial_Searching_Individuals();

virtual~Artificial_Searching_Individuals();

};
```

Suppose r_1, r_2, r_3 are three random real numbers, $0 \le r_1, r_2, r_3 \le 1$, the individual's position is X_i in i-th iteration, the historical best solution of the individual is X_s, the historical best solution of the swarm is X_g, the location of the called individual which find the better goal is X_{call}, the distance function is $\|\cdot\|$, and the absolute value function is $|\cdot|$, then the three behaviors can be described as follows:

void AS_synergetic()

{

$$X_{i+1} = X_{call} + r_1 * AS\_step * \frac{X_{call} - X_i}{\|X_{call} - X_i\|};$$

};

void AS_reconnaissance()

{

$$X = X_i + r_1 * |X_s - X_i| + r_2 * |X_g - X_i|;$$

$$X_{i+1} = X + r_3 * AS\_step * \frac{X - X_i}{\|X - X_i\|};$$

}.

void AS_move ()

{

$$X_{i+1} = X_i + r_1 * AS\_step;$$

};

4. SIMULATION AND PERFORMANCE ANALYSIS

4.1. Simulation Functions

$$F_1(x,y)=\frac{\sin(x)}{x}\times\frac{\sin(y)}{y} \quad x,y\in[-10,10] \tag{1}$$

$$F_2(X)=\sum_{i=1}^{n-1}100\left(x_i^2-x_{i+1}\right)^2+\left(1-x_i\right)^2$$
$$x_i\in[-2.048,2.048],\ i=1,2,\cdots,n \tag{2}$$

$$F_3(x_1,x_2,x_3)=30+\sum_{i=1}^{3}\left(\frac{1}{10}(x_i-20)^2-9\cos\frac{2\pi}{5}x_i\right)$$
$$x_i\in[1,39] \quad i=1,2,3 \tag{3}$$

$$F_4(x_1,x_2)=\left[1+(x_1+x_2+)^2(19-14x_1+3x_1^2-14x_2\right.$$
$$+6x_1x_2+3x_2^2)\left[30+(2x_1-3x_2)^2(18-32x_1\right.$$
$$\left.\left.+12x_1^2+48x_2-36x_1x_2+27x_2^2)\right]\right.$$
$$x_i\in[-2,2] \quad i=1,2 \tag{4}$$

$$F_5(X)=-20\exp\left(-0.2\sqrt{\frac{\sum_{i=1}^{n}x_i^2}{n}}\right)-\exp\left(\frac{\sum_{i=1}^{n}\cos(2\pi x_i)}{n}\right)$$
$$+20+e$$
$$x_i\in[-32,32],i=1,2,\cdots,n \tag{5}$$

$$F_6(X)=\sum_{i=1}^{n}\frac{x_i^2}{4000}-\prod_{i=1}^{n}\left(\cos\left(\frac{x_i}{\sqrt{i}}\right)\right)+1$$
$$x_i\in[-600,600],i=1,2,\cdots,n \tag{6}$$

$$F_7(X) = 10n + \sum_{i=1}^{n}\left(x_i^2 - 10\cos(2\pi x_i)\right)$$
$$x_i \in [-10, 10], i = 1, 2, \cdots, n \tag{7}$$

$$F_8(X) = \sum_{i=1}^{n}\left(\lfloor x_i + 0.5 \rfloor\right)^2$$
$$x_i \in [-100, 100], i = 1, 2, \cdots, n \tag{8}$$

$$F_9(x_1, x_2) = (x_1 - 10)^3 + (x_2 - 20)^3$$
$$s.t. \begin{cases} g_1(x_1, x_2) = 100 - (x_1 - 5)^2 - (x_2 - 5)^2 \le 0 \\ g_2(x_1, x_2) = -82.81 + (x_1 - 6)^2 - (x_2 - 5)^2 \le 0 \\ 13 \le x_1 \le 100, 0 \le x_2 \le 100 \end{cases} \tag{9}$$

$$F_{10}(x_1, x_2) = \left(x_1^2 + x_2 - 11\right)^2 + \left(x_1 + x_2^2 - 7\right)^2$$
$$s.t. \begin{cases} g_1(x_1, x_2) = (x_1 - 0.05)^2 + (x_2 - 2.5)^2 - 4.84 \le 0 \\ g_2(x_1, x_2) = 4.84 - x_1^2 - (x_2 - 2.5)^2 \le 0 \\ 0 \le x_1, x_2 \le 6 \end{cases} \tag{10}$$

F_1 has a global maximum 1 at point [0, 0], and a lot of local maximums distributed around it. The usual algorithms are easy to fall into local maximums or vibrate between the local minimum and maximum. It is often chosen to verify the validity of the algorithm.

F_2 is the general Rosenbrock function. It has the global minimum value 0 at the points [1, 1, ⋯, 1]. As the global optimum is inside a long, narrow, parabolic shaped flat valley, it is difficult for global optimization, and often be used to detect the searching ability of algorithm.

F_3 has a global minimum value 3 at point [20, 20, 20], and a lot of local minimum value distributed around it, it is a multi-variable function.

F_4 is the Goldstein-Price function and has global minimum 3 at point [0, −1].

F_5 is the Ackley's Path function. It is a typical complex multidimensional multimodal function. It has the global minimum 0 at point [0, 0, ⋯, 0].

F_6 is the Griewank function. It is a typical nonlinear multimodal function and has the extensive search space. It has the global minimum 0 at point [0, 0, ···, 0] and a lot of local minimums distributed around it. The number of local minimums relates to function's dimension.

F_7 is the Rastrigrin function. It has the global minimum 0 at point [0, 0, ···, 0].

F_8 is the Step function. It has the global minimum 0 at point [0, 0, ···, 0].

F_9 is a constrained optimization problem; it subjects to some constrained conditions. It has the minimum value −6961.81388 at point (14.095, 0.84296).

F_{10} is a constrained optimization problem. It has the minimum value 13.59084 at point (2.2468, 2.3818).

4.2. Performance Analysis and Comparison

ASSA is used to solve function F_1. The size of the searching swarm is 10, searching step is 0.3, Pc is 0.006, and the number of iterations is 100. **Figure 1** shows the initial distribution of searching swarm, the searching individuals distribute randomly in the region.

After 20 iterations, the distribution of swarm is shown in **Figure 2**. Some individuals have searched near the global optimum position.

After 50 iterations, the swarm distribution is shown in **Figure 3**. More individuals have searched near the global optimal solution.

As **Figure 4** shown, after 100 iterations, most of the individuals have searched near the optimal solution, and the value of the best individual is getting closer and closer to the optimal value, that concretely shows the effecttiveness of the algorithm.

In order to describe the process of the searching swarm hunting for the optimal solution concretely,**Figure 5** shows the change track of the best individual's fitness value intercepted from the 34th to 100th iteration. ASSA shows a powerful searching ability, and goes forward to the optimal solution all along. The value of the best individual is more and more approaching to the optimal value that reflects a better searching ability and searching efficiency of the ASSA.

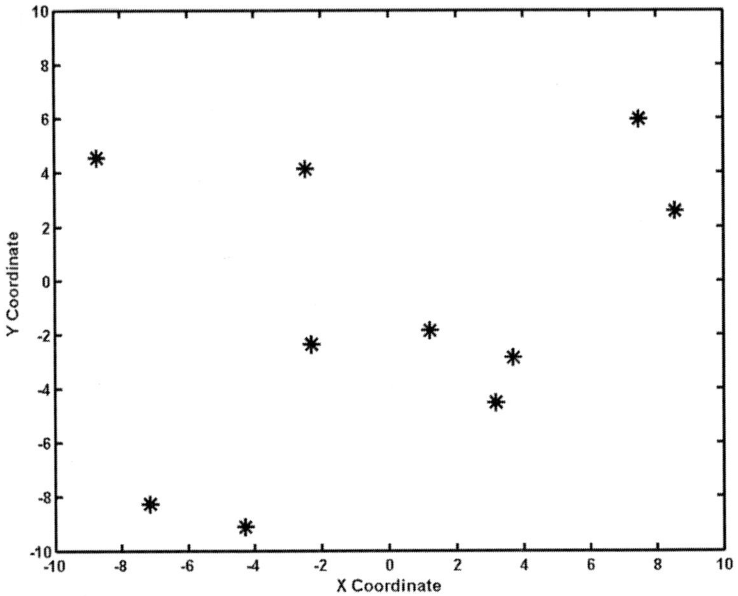

Figure 1. Distribution of the initial swarm.

ASSA and GA are used respectively to solve function F_2. Select n is 3, the swarm size is 10, searching step is 0.2, Pc is 0.002, and the iteration is 100. The GA uses binary code, the crossover probability is 0.6, mutation probability is 0.001, and the length of binary code is 30. The test results are shown in **Figure 6**. In the early running time, because of the restriction of the searching step, the ASSA is slower than the GA, but ASSA exceeds the GA and gets the better optimization results finally.

In order to compare the performance of ASSA and PSO, function F_4 is selected to do simulation experiment. The swarm size is 10, Pc is 0.02, iteration is 100, searching step of ASSA is 0.6, the inertia weight of PSO is descending, experiments are 100 times, and the results are shown in**Figure 7**. ASSA has better performance than PSO besides several running cases.

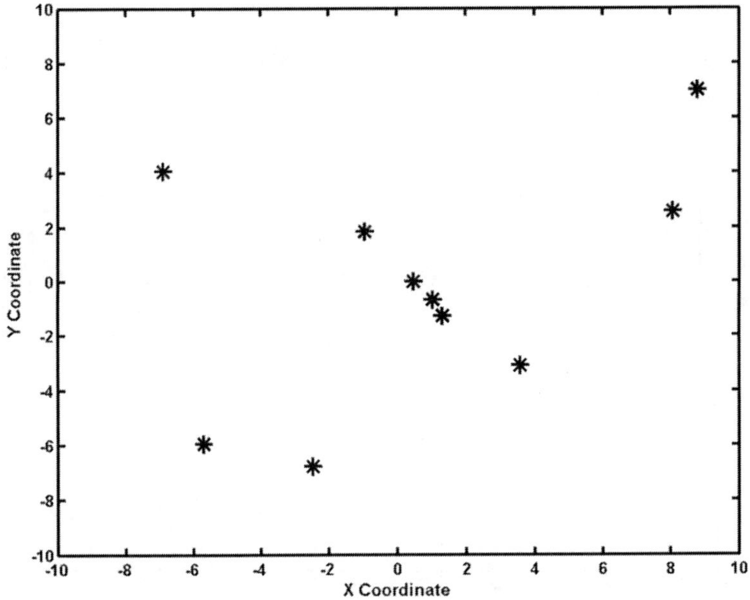

Figure 2. Distribution of the swarm after 20 iterations.

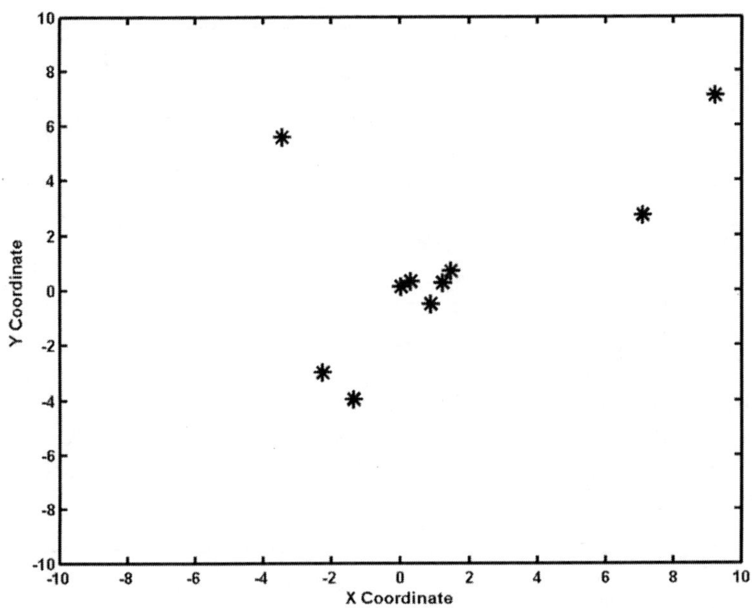

Figure 3. Distribution of the swarm after 50 iterations.

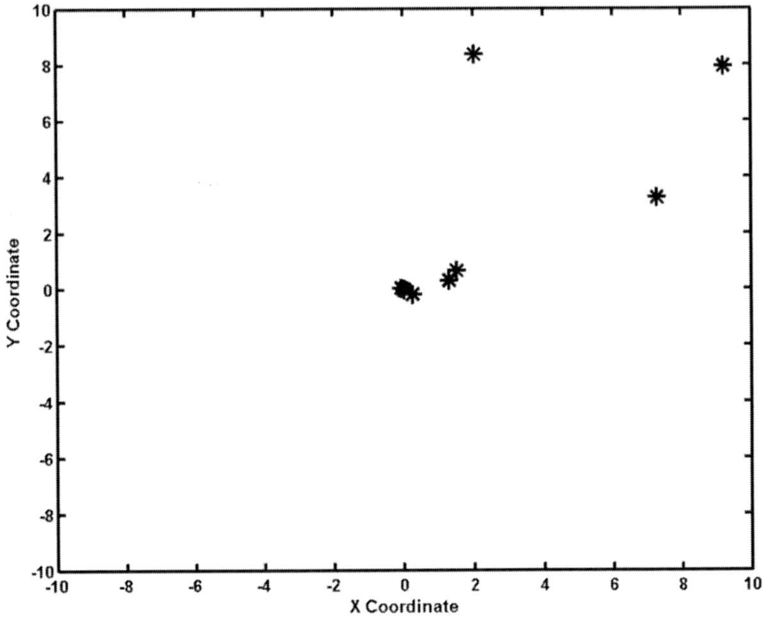

Figure 4. Distribution of the swarm after 100 iterations.

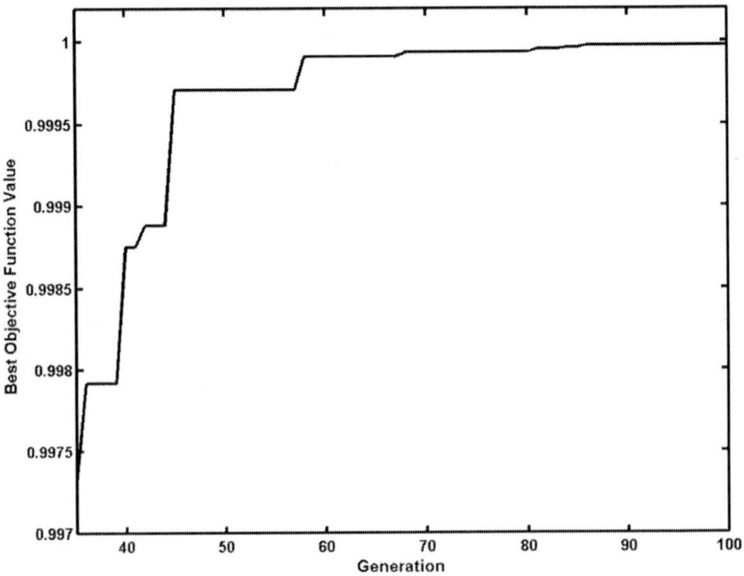

Figure 5. Latter part of the objective function value of the best individual.

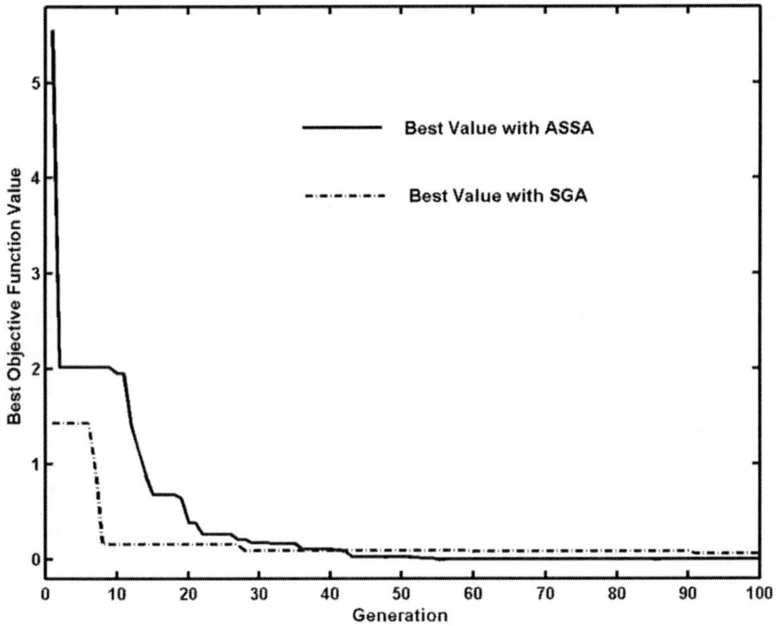

Figure 6. Comparison of the best individual value of ASSA and SGA.

AFSA is a new intelligent optimization algorithm based on the simulation of the fish behaviors. Function F_3 is selected to do simulative experiment with ASSA and AFSA respectively. The Swarm size is 100, Pc is 0.006, iteration is 100, searching step of ASSA and AFSA are 0.6, the fish vision of AFSA is 15, and experiments are 100 times. The results are shown as **Figure 8**. ASSA has better holistic performance than AFSA.

For testing the searching ability of ASSA in high dimension space function F_5, F_6, F_7 and F_8 are selected to do experiment respectively. The dimension is 20, swarm size is 30, the number of iteration is 50, experiments are 50 times, searching step of F_5 is 9, searching step of F_6 is 150, searching step of F_7 is 5.5 searching step of F_8 is 40, Pc is 0.006, and the results are shown as**Table 1**.

The change track of the best individual's fitness value of F_5, F_6 are shown as Figures 9 and 10 respectively.

From the results ASSA has the better performance in high dimension optimization, but does not behaves outstanding ability in running time. Having more move rules may be the main reason.

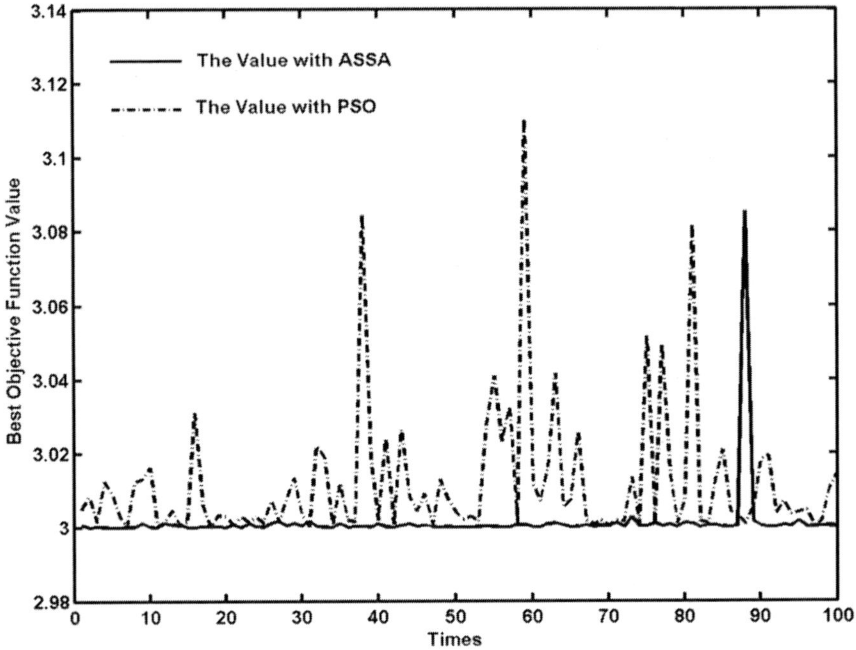

Figure 7. Comparison of the best individual value of ASSA and PSO.

4.3. Experiments for Solving Constrained Optimization Problems

For solving the constraint optimization problem, Penalty Function Method (PFM) is a useful method. Penalty function method converts the object function into the penalty function which included the constrained information, and the process of solving constraint optimization problem is transformed to the process of continuously solving unrestraint optimization problem.

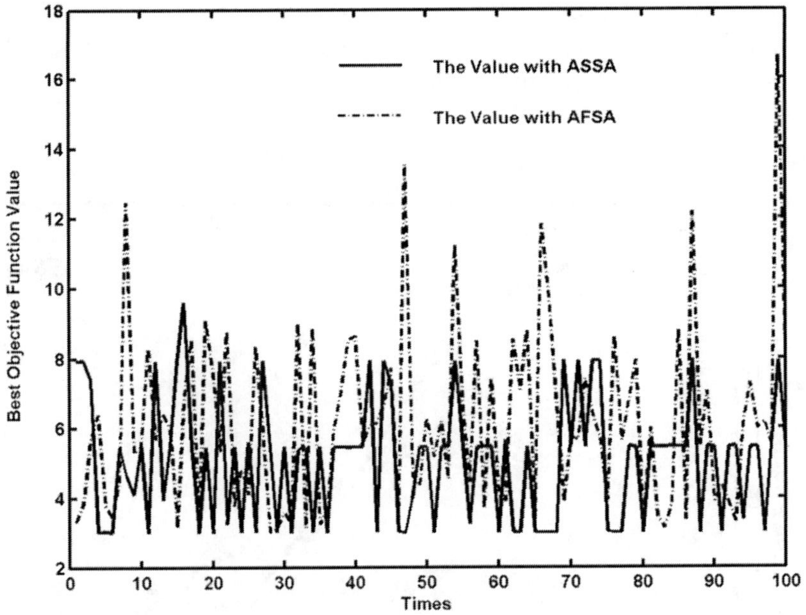

Figure 8. Comparison of the best individual value of ASSA and AFSA.

Table 1. Optimal value of algorithms.

Function	ASSA	AFSA	GA	PSO
F5	3.547	8.396	14.675	5.004
F6	1.475	8.881	39.784	1.997
F7	37.285	124.47	174.71	95.197
F8	47.122	2462.1	8959.7	227.13
CPU time (s)	2.982	7.566	2.2	1.326

Note: The data of GA are from reference [9].

Penalty function method has the penalty factor sequence which makes difficulty to program. The penalty factor (has fixed value) usually be used for simple.

About function F_9, with PFM, select swarm size is 100, the number of iteration is 500, searching step = 0.6, experiment times = 10, and the penalty factor = 10^8. The results are shown as **Table 2**.

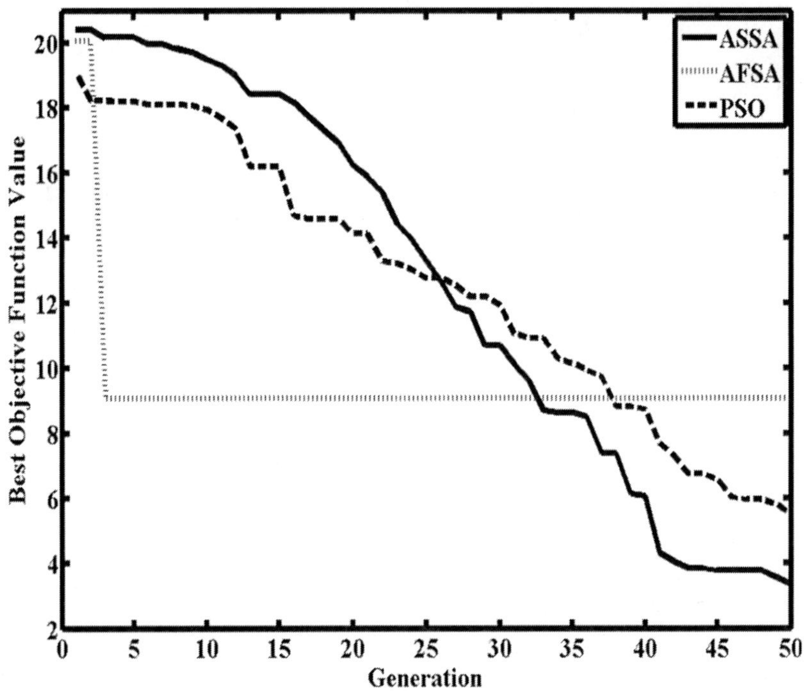

Figure 9. Comparison of the best individual value of F_5.

About function F_{10}, with PFM, select swarm size is 200, the number of iteration is 500, searching step = 0.2, experiment times = 10, and the penalty factor = 10^8. The results are shown as **Table 3**.

From simulative results ASSA is an effective algorithm for solving constrained optimization problems.

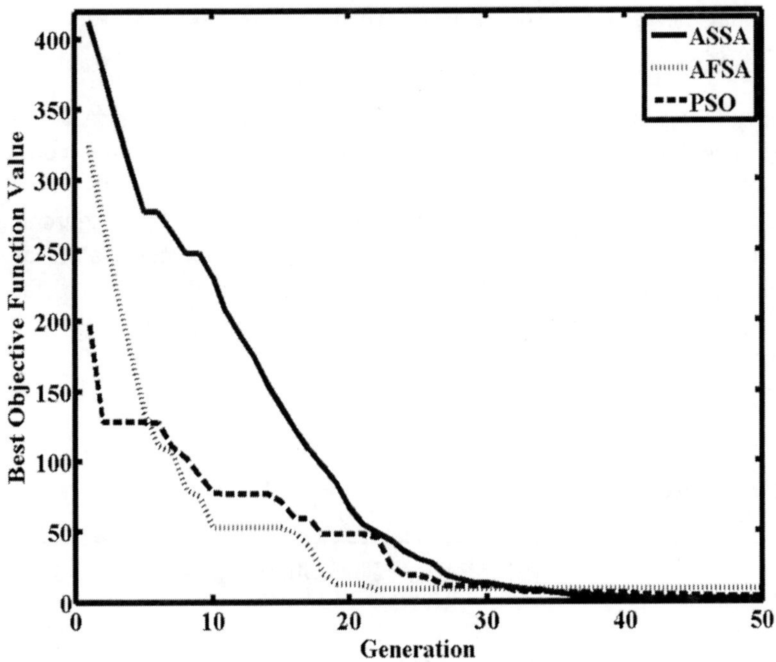

Figure 10. Comparison of the best individual value of F_6.

4.4. Parameter Analysis and Comparison

According to the synergetic move rule, the value of synergetic parameter is a key factor that restrains local convergence and influences the performance of ASSA.

For verifying the analysis above, function F_1 is selected to do the simulation tests. The swarm size is 10, searching step is 0.3, the number of iteration is 100, and the results are shown as **Figure 11**.

From the results selecting the value of synergetic parameter properly can enhance the performance of ASSA. The value of the probability is suggested between 0.001 - 0.1.

Searching step is one of the main factors which affect the performance of ASSA. Select function F_3, the population size is 100, Pc is 0.006, the half length of the searching area as the initial vale of the searching step, which gradually decreases as the algorithm running. The results are shown as **Figure 12**.

From the results, the larger searching step the swarm move faster in the early running of the algorithm, but the algorithm losses the continual searching capability in the latter running time; if searching step is smaller, the moving of the swarm is slower, although it can continue searching in the latter running time, it can't get better results for constraints of the iterative number. According to the simulation results, the more satisfactory results of searching step are between 15% - 30% length of the searching area.

Table 2. Results of function F_9.

Method	The optimal value		
	Best	*Median*	*Worst*
Reference [11]	−6961.814	−6955.812	−6954.446
ASSA	−6961.801	−6954.734	−6951.622

Table 3. Results of function F_{10}.

Method	The optimal value		
	Best	*Median*	*Worst*
Reference [10]	13.59084	13.59087	13.59097
Reference [11]	13.59084	13.59084	13.59085
ASSA	13.59089	13.59108	13.59144

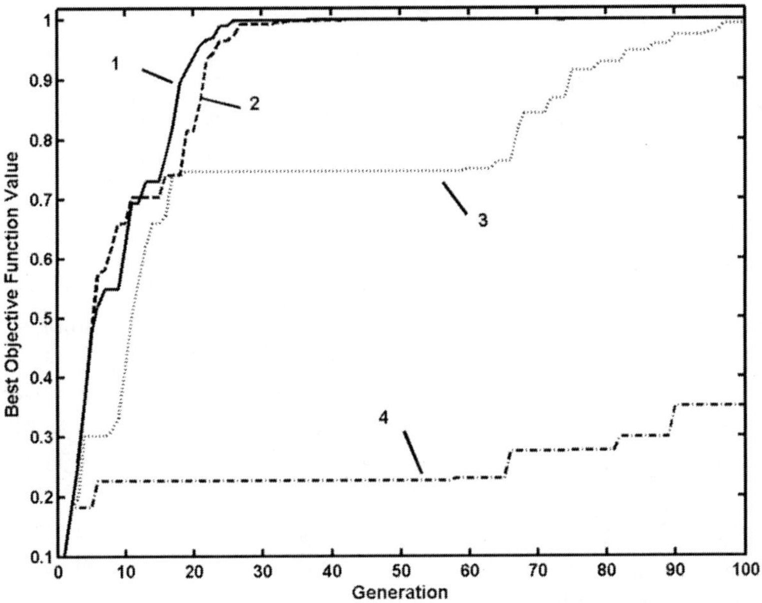

Figure 11. Comparative results with different value of Pc (1: Pc = 0.002, 2: Pc = 0.02, 3: Pc = 0.2, 4: Pc = 1.0).

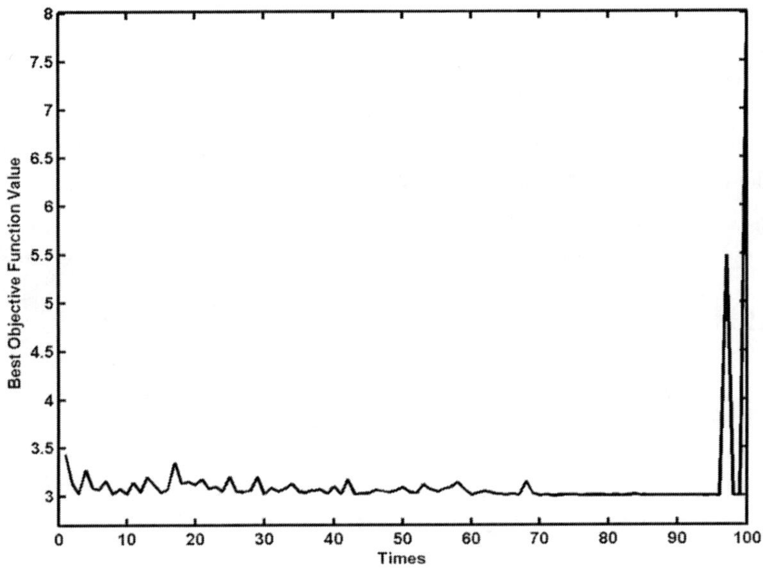

Figure 12. Influence of searching step.

As well known the population size is a key factor influenced the performance of BIOA. In general, the larger the population, the better the performance, but it does not meet the linear relationship. For ASSA the advisable value of the population size is suggested between 10 - 200.

5. CONCLUSIONS

Artificial Searching Swarm Algorithm (ASSA) is based on the simulation of the operation principle and abstract model of BIOA. In ASSA, the process of solving optimization problem is transformed into the process of searching optimal goal by searching swarm with set rules, and the optimal solution is found through the search and cooperation of the searching individuals.

ASSA possesses easy programmable, simple structure, insensitive to initial values, strong global searching ability, and small swarm running ability character. The algorithm is used for optimizing multivariable functions and the results produced by ASSA, Genetic Algorithm (GA), Particle Swarm Optimization (PSO), Artificial FishSwarm Algorithm (AFSA) have been compared. The experiment results showed that ASSA is an effective optimization algorithm.

ACKNOWLEDGEMENTS

This work was supported by the Natural Science Foundation of Hebei Province under Grant No. E2011202040.

REFERENCES

1. J. H Holland "Adaptation in Nature and Artificial System," MIT Press, Cambridge, 1992.

2. M. Dorigo, "Optimization, Learning and Natural Algorithms," Ph.D. Thesis, Department of Electronics, Polytechnic of Milan, Milan, 1992.

3. J. Kennedy and R. C. Eberhart, “Particle Swarm Optimization,” Proceedings of the IEEE International Conference on Neural Networks, Perth, 27 November-1 December 1995, pp. 1942-1948.

4. X. L. Li, Z. J. Shao and J. X. Qian, “An Optimization Method Based on Autonomous Animats: Fish-Swarm Algorithm,” Systems Engineering-Theory & Practice, Vol. 22, No. 11, 2002, pp. 32-38.

5. M. Eusuffm and K. E. Lansey, “Optimization of Water Distribution Network Design Using Shuffled Frog Leaping Algorithm,” Journal of Water Resources Planning and Management, Vol. 129, No. 3, 2003, pp. 210-225.

6. T. G. Chen, “A Simulative Bionic Intelligent Optimization Algorithm: Artificial Searing Swarm Algorithm and its performance Analysis,” Proceedings of the Second International Joint Conference on Computational Sciences and Optimization, Sanya, Hainan, 24-26 April 2009, Vol. 2, pp. 864-866.

7. H. B. Duan, D. B. Wang and X. F. Yu, “Research on some Novel Bionic Optimization Algorithms,” Computer Simulation, Vol. 24, No. 3, 2007, pp.169-172.

8. H. Wang and F. Qian, “Swarm Intelligent Optimization Algorithms,” Chemical Automation and Instrumentation, Vol. 34, No. 5, 2007, pp.7-13.

9. H. P. Ma, H. Li and X. Y. Run, “Species Migration-Based Optimization Algorithm and Performance Analysis,” Control Theory & Applications, Vol. 27, No. 3, 2010, pp. 329-334.

10. X. H. Wang, X. M. Zheng and J. M. Xiao, “Artificial Fish-Swarm Algorithm for Solving Constrained Optimization Problems,” Computer Engineering and Applications, Vol. 43, No. 3, 2007, pp. 40-42.

11. Y. R. Zhou, J. X. Zhou and Y. Wang, “An Optimised Evolutionary Algorithms for Nonparameter Penalty Function,” Computer Engineering, Vol. 31, No. 10, 2005, pp. 31-33.

Chapter 6

Firefly algorithm: Recent advances and applications

Xin-She Yang1,*, Xingshi He2

1 School of Science and Technology, Middlesex University, The Burroughs, London NW4 4BT, UK

2 School of Science, Xi'an Polytechnic University, No. 19 Jinhua South Road, Xi'an 710048, China

ABSTRACT

Nature-inspired metaheuristic algorithms, especially those based on swarm intelligence, have attracted much attention in the last ten years. Firefly algorithm appeared in about five years ago, its literature has expanded dramatically with diverse applications. In this paper, we will briefly review the fundamentals of firefly algorithm together with a selection of recent publications. Then, we discuss the optimality associated with balancing exploration and exploitation, which is essential for all metaheuristic algorithms. By comparing with intermittent search strategy, we conclude that metaheuristics such as firefly algorithm are better than the optimal intermittent search strategy. We also analyse algorithms and their implications for higherdimensional optimisation problems.

KEYWORDS

algorithms, bat algorithm, cuckoo search, firefly algorithm, metaheuristic, nature-inspired algorithms

1 INTRODUCTION

Metaheuristic algorithms form an important part of contemporary global optimisation algorithms, computational intelligence and soft computing. These algorithms are usually nature-inspired with multiple interacting agents. A subset of metaheuristcs are often referred to as swarm intelligence (SI)-based algorithms, and these SI-based algorithms have been developed by mimicking the so-called SI characteristics of biological agents such as birds, fish, humans and others. For example, particle swarm optimisation was based on the swarming behaviour of birds and fish (Kennedy and Eberhar, 1995), while the firefly algorithm (FA) was based on the flashing pattern of tropical fireflies (Yang, 2008, 2009) and cuckoo search algorithm was inspired by the brood parasitism of some cuckoo species (Yang and Deb, 2009).

In the last two decades, more than a dozen new algorithms such as particle swarm optimisation, differential evolution, bat algorithm, FA and cuckoo search have appeared and they have shown great potential in solving tough engineering optimisation problems (Yang, 2008, 2010a, 2010b; Bansal and Deep, 2008; Floudas and Pardolos, 2009; Parpinelli and Lopes, 2011; Gandomi et al., 2011). Among these new algorithms, it has been shown that FA is very efficient in dealing with multimodal, global optimisation problems.

In this paper, we will first outline the fundamentals of FA, and then review the latest developments concerning FA and its variants. We also highlight the reasons why FA is so efficient. Furthermore, as the balance of exploration and exploitation is important to all metaheuristic algorithms, we will then discuss the optimality related to search landscape and algorithms. Using the intermittent search strategy and numerical experiments, we show that FA is significantly more efficient than intermittent search strategy.

2 FA AND COMPLEXITY

2.1 Firefly algorithm

FA was first developed by Xin-She Yang in late 2007 and 2008 at Cambridge University (Yang, 2008, 2009), which was based on the flashing patterns and behaviour of fireflies. In essence, FA uses the following three idealised rules:

- Fireflies are unisex so that one firefly will be attracted to other fireflies regardless of their sex.
- The attractiveness is proportional to the brightness, and they both decrease as their distance increases. Thus for any two flashing fireflies, the less brighter one will move towards the brighter one. If there is no brighter one than a particular firefly, it will move randomly.
- The brightness of a firefly is determined by the landscape of the objective function.

As a firefly's attractiveness is proportional to the light intensity seen by adjacent fireflies, we can now define the variation of attractiveness β with the distance r by

$$\beta = \beta_0 e^{-\gamma r^2},$$

where β_0 is the attractiveness at $r = 0$.

The movement of a firefly i is attracted to another more attractive (brighter) firefly j is determined by

$$x_i^{t+1} = x_i^t + \beta_0 e^{-\gamma r_{ij}^2} \left(x_j^t - x_i^t\right) + \alpha_t \, \epsilon_i^t,$$

where the second term is due to the attraction. The third term is randomisation with α_t being the randomisation parameter, and ϵ_i^t is a vector of random numbers drawn from a Gaussian distribution or uniform distribution at time t. If $\beta_0 = 0$, it becomes a simple random walk. On the other hand, if $\gamma = 0$, it reduces to a variant of particle swarm optimisation (Yang, 2008). Furthermore, the randomisation ϵ_i^t can easily be extended to other distributions such as Lévy flights (Yang, 2008). A demo version of FA implementation by Xin-She Yang, without Lévy flights for simplicity, can be found at Mathworks file exchange website (http://www.mathworks.com/matlabcentral/fileexchange/29693-firefly-algorithm).

2.2 Parameter settings

As α_t essentially control the randomness (or, to some extent, the diversity of solutions), we can tune this parameter during iterations so that it can vary with the iteration counter t. So a good way to express α_t is to use

$$\alpha_t = \alpha_0 \delta^t, \quad 0 < \delta < 1),$$

where α_0 is the initial randomness scaling factor, and δ is essentially a cooling factor. For most applications, we can use δ = 0:95 to 0:97 (Yang, 2008).

Regarding the initial α_0, simulations show that FA will be more efficient if α_0 is associated with the scalings of design variables. Let L be the average scale of the problem of interest, we can set $\alpha_0 = 0.01L$ initially. The factor 0:01 comes from the fact that random walks requires a number of steps to reach the target while balancing the local exploitation without jumping too far in a few steps (Yang, 2009, 2010a).

The parameter β controls the attractiveness, and parametric studies suggest that $\beta_0 = 1$ can be used for most applications. However, γ should be also related to the scaling L. In general, we can set $\gamma = 1/\sqrt{L}$. If the scaling variations are not significant, then we can set $\gamma = O(1)$.

For most applications, we can use the population size n = 15 to 100, though the best range is n = 25 to 40 (Yang, 2008, 2009).

2.3 Algorithm complexity

Almost all metaheuristic algorithms are simple in terms of complexity, and thus they are easy to implement. FA has two inner loops when going through the population n, and one outer loop for iteration t. So the complexity at the extreme case is $O(n^2t)$. As n is small (typically, $n = 40$), and

t is large (say, t = 5,000), the computation cost is relatively inexpensive because the algorithm complexity is linear in terms of t. The main computational cost will be in the evaluations of objective functions, especially for external black-box type objectives. This latter case is also true for all metaheuristic algorithms. After all, for all optimisation problems, the most computationally extensive part is objective evaluations.

If n is relatively large, it is possible to use one inner loop by ranking the attractiveness or brightness of all fireflies using sorting algorithms. In this case, the algorithm complexity of FA will be $O(nt\log(n))$.

2.4 FA in applications

FA has attracted much attention and has been applied to many applications (Apostolopoulos and Vlachos, 2011; Chatterjee et al., 2012; Hassanzadeh et al., 2011; Sayadi et al., 2010; Yang, 2010c; Horng et al., 2012; Horng, 2012). Horng et al. (2012) and Horng (2012) demonstrated that firefly-based algorithm used least computation time for digital image compression, while Banati and Bajaj (2011) used FA for feature selection and showed that FA produced consistent and better performance in terms of time and optimality than other algorithms.

In the engineering design problems, Gandomi et al. (2011) and Azad and Azad (2011) confirmed that FA can efficiently solve highly non-linear, multimodal design problems. Basu and Mahanti (2011) as well as Chatterjee et al. (2012) have applied FA in antenna design optimisation and showed that FA can outperform artificial bee colony (ABC) algorithm. In addition, Zaman and Matin (2012) have also found that FA can outperform PSO and obtained global best results.

Sayadi et al. (2010) developed a discrete version of FA which can efficiently solve NP-hard scheduling problems, while a detailed analysis has demonstrated the efficiency of FA over a wide range of test problems, including multobjective load dispatch problems (Apostolopoulos and Vlachos, 2011; Yang, 2009, 2012a). Furthermore, FA can also solve scheduling and travelling salesman problem in a promising way (Palit et al., 2011; Jati and Suyanto, 2011; Yousif et al., 2011).

Classifications and clustering are another important area of applications of FA with excellent performance (Senthilnath et al., 2011; Rajini and David,

2012). For example, Senthilnath et al. (2011) provided an extensive performance study by compared FA with 11 different algorithms and concluded that FA can be efficiently used for clustering. In most cases, FA outperform all other 11 algorithms. In addition, FA has also been applied to train neural networks (Nandy et al., 2012).

For optimisation in dynamic environments, FA can also be very efficient as shown by Farahani et al. (2011a, 2011b) and Abshouri et al. (2011).

2.5 Variants of FA

For discrete problems and combinatorial optimisation, discrete versions of FA have been developed with superior performance (Sayadi et al., 2010; Hassanzadeh et al., 2011; Jati and Suyanto, 2011; Fister et al., 2012; Durkota, 2011), which can be used for travelling-salesman problems, graph colouring and other applications.

In addition, extension of FA to multiobjective optimisation has also been investigated (Apostolopoulos and Vlachos, 2011; Yang, 2012b).

A few studies show that chaos can enhance the performance of FA (Coelho et al., 2011; Yang, 2011), while other studies have attempted to hybridise FA with other algorithms to enhance their performance (Giannakouris et al., 2010; Horng, 2012; Horng and Liou, 2011; Rampriya et al., 2010).

3. WHY FA IS SO EFFICIENT?

As the literature about FA expands and new variants have emerged, all pointed out the FA can outperform many other algorithms. Now we may ask naturally 'Why is it so efficient?'. To answer this question, let us briefly analyse the FA itself.

FA is swarm-intelligence-based, so it has the similar advantages that other swarm-intelligence-based algorithms have. In fact, a simple analysis of parameters suggests that some PSO variants such as accelerated PSO (Yang et al., 2011) are a special case of FA when $\gamma = 0$ (Yang, 2008).

However, FA has two major advantages over other algorithms: automatical subdivision and the ability of dealing with multimodality. First, FA is based on attraction and attractiveness decreases with distance. This leads to the fact that the whole population can automatically subdivide into subgroups, and each group can swarm around each mode or local optimum. Among all these modes, the best global solution can be found. Second, this subdivision allows the fireflies to be able to find all optima simultaneously if the population size is sufficiently higher than the number of modes. Mathematically, $1/\sqrt{\gamma}$ controls the average distance of a group of fireflies that can be seen by adjacent groups. Therefore, a whole population can subdivide into subgroups with a given, average distance. In the extreme case when $\gamma = 0$, the whole population will not subdivide. This automatic subdivision ability makes it particularly suitable for highly non-linear, multimodal optimisation problems.

In addition, the parameters in FA can be tuned to control the randomness as iterations proceed, so that convergence can also be sped up by tuning these parameters. These above advantages makes it flexible to deal with continuous problems, clustering and classifications, and combinatorial optimisation as well.

As an example, let use two functions to demonstrate the computational cost saved by FA. For details, please see the more extensive studies by Yang (2009). For De Jong's function with d = 256 dimensions

$$f(\boldsymbol{x}) = \sum_{i=1}^{256} x_i^2.$$

Genetic algorithms required 25,412 ± 1,237 evaluations to get an accuracy of 10^{-5} of the optimal solution, while PSO needed 17,040 ± 1,123 evaluations. For FA, we achieved the same accuracy by 5,657 ± 730. This save about 78% and 67% computational cost, compared to GA and PSO, respectively.

For Yang's forest function

$$f(\boldsymbol{x}) = \left(\sum_{i=1}^{d} |x_i|\right) \exp\left[-\sum_{i=1}^{d} \sin\left(x_i^2\right)\right], \quad -2\pi \leq x_i \leq 2\pi,$$

GA required 37,079 ± 8,920 with a success rate of 88% for d = 16, and PSO required 19,725 ± 3,204 with a success rate of 98%. FA obtained a 100% success rate with just 5,152 ± 2,493. Compared with GA and PSO, FA saved about 86% and 74%, respectively, of overall computational efforts.

As an example for automatic subdivision, we now use the FA to find the global maxima of the following function

$$f(x) = \left(\sum_{i=1}^{d} |x_i| \right) \exp\left(-\sum_{i=1}^{d} x_i^2 \right),$$

with the domain $-10 \le x_i \le 10$ for all (i = 1, 2, ..., d) where d is the number of dimensions. This function has multiple global optima (Yang, 2010c). In the case of d = 2, we have 4 equal maxima $f_* = 1/\sqrt{e} \approx 0.6065$ at (1/2, 1/2), (1/2, −1/2), (−1/2, 1/2) and (−1/2, −1/2) and a unique global minimum at (0, 0).

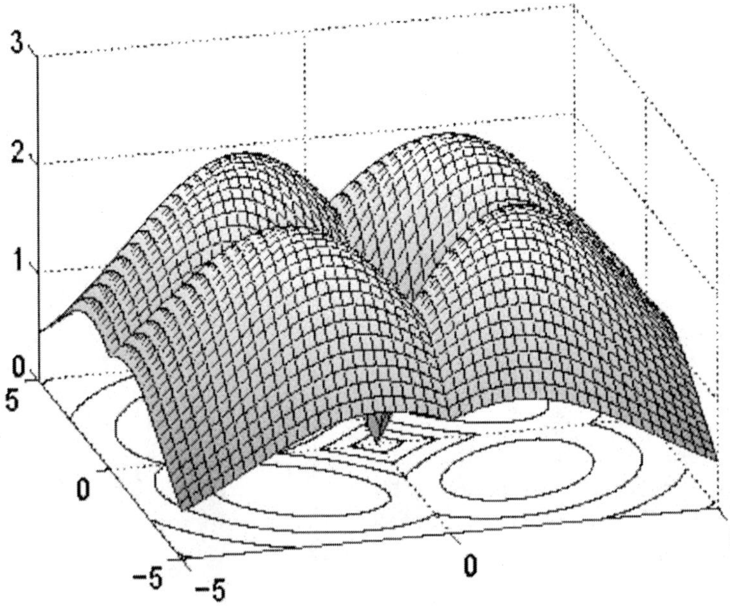

Figure 1 Four global maxima at (±1/2;±1/2)

In the 2D case, we have a four-peak function are shown in Figure 1, and these global maxima can be found using the implemented FAs after about

500 function evaluations. This corresponds to 25 fireflies evolving for 20 generations or iterations. The initial locations of 25 fireflies and their final locations after 20 iterations are shown in Figure 2.

We can see that FA can indeed automatically subdivide its population into subgroups.

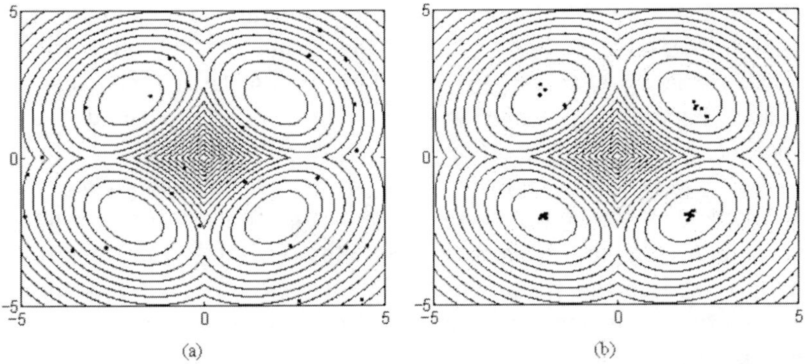

Figure 2 (a) Initial locations of 25 fireflies and (b) their final locations after 20 iterations

4. SEARCH OPTIMALITY

4.1 Intensification versus diversification

The main components of any metaheuristic algorithms are: intensification and diversification, or exploitation and exploration (Blum and Roli, 2003; Yang, 2012a). Diversification means to generate diverse solutions so as to explore the search space on the global scale, while intensification means to focus on the search in a local region by exploiting the information that a current good solution is found in this region. This is in combination with the selection of the best solutions.

Exploration in metaheuristics can be achieved often by the use of randomisation (Blum and Roli, 2003; Yang, 2008, 2009), which enables an algorithm to have the ability to jump out of any local optimum so as to explore the search globally. Randomisation can also be used for local search around the current best if steps are limited to a local region. When the steps are large, randomisation can explore the search space on a global scale. Fine-tuning the right amount of randomness and balancing local search and global search are crucially important in controlling the performance of any metaheuristic algorithm.

Exploitation is the use of local knowledge of the search and solutions found so far so that new search moves can concentrate on the local regions or neighbourhood where the optimality may be close; however, this local optimum may not be the global optimality. Exploitation tends to use strong local information such as gradients, the shape of the mode such as convexity, and the history of the search process. A classic technique is the so-called hill-climbing which uses the local gradients or derivatives intensively.

Empirical knowledge from observations and simulations of the convergence behaviour of common optimisation algorithms suggests that exploitation tends to increase the speed of convergence, while exploration tends to decrease the convergence rate of the algorithm. On the other hand, too much exploration increases the probability of finding the global optimality, while strong exploitation tends to make the algorithm being trapped in a local optimum. Therefore, there is a fine balance between the right amount of exploration and the right degree of exploitation. Despite its importance, there is no practical guideline for this balance.

4.2 Landscape-based optimality or algorithm optimality?

It is worth pointing out that the balance for exploration and exploitation was often discussed in the context of optimisation algorithms; however, this can be more often related to the problem type of interest, and thus such balance is problem-specific, depending on the actual landscape of the objective function. Consequently, there may be no universality at all. Therefore, we may have to distinguish landscape-dependent optimality for exploration and exploitation, and algorithm-based optimality. The former focuses on landscape, while the latter focuses on algorithms.

Landscape-based optimality focuses on the information of the search landscape and it is hoped that a better (or even best) approach/algorithm may find the optimal solutions with the minimum effort (time, number of evaluations), while algorithm-based approach treats objective functions as a black-box type and tries to use partly the available information during iterations to work out the best ways to find optimal solutions. As to which approach is better, the answer may depend on one's viewpoint and focus. In any case, a good combination of both approaches may be needed to reach certain optimality.

4.3 Intermittent search strategy

Even there is no guideline in practice for landscape-based optimality, some preliminary work on the very limited cases exists in the literature and may provide some insight into the possible choice of parameters so as to balance these components. In fact, intermittent search strategy is a landscape-based search strategy (Bénichou et al., 2006).

Intermittent search strategies concern an iterative strategy consisting of a slow phase and a fast phase (Bénichou et al., 2006, 2011). Here the slow phase is the detection phase by slowing down and intensive, static local search techniques, while the fast phase is the search without detection and can be considered as an exploration technique. For example, the static target detection with a small region of radius *a* in a much larger region *b* where $a \ll b$ can be modelled as a slow diffusive process in terms of random walks with a diffusion coefficient *D*.

Let τ_a and τ_b be the mean times spent in intensive detection stage and the time spent in the exploration stage, respectively, in the 2D case. The diffusive search process is governed by the mean first-passage time satisfying the following equations (Bénichou et al., 2011)

$$D\nabla_r^2 t_1 + \frac{1}{2\pi\tau_a}\int_0^{2\pi}[t_2(r) - t_1(r)]d\theta + 1 = 0,$$

$$u \cdot \nabla_r t_2(r) - \frac{1}{\tau_b}[r_2(r) - t_1(r)] + 1 = 0,$$

where t_2 and t_1 are mean first-passage times during the search process, starting from slow and fast stages, respectively, and u is the mean search speed.

After some lengthy mathematical analysis, the optimal balance of these two stages can be estimated as

$$r_{\text{optimal}} = \frac{\tau_a}{\tau_b^2} \approx \frac{D}{a^2} \frac{1}{\left[2 - \frac{1}{\ln(b/a)}\right]^2}.$$

Assuming that the search steps have a uniform velocity *u* at each step on average, the minimum times required for each phase can be estimated as

$$\tau_a^{\min} \approx \frac{D}{2u^2} \frac{\ln^2(b/a)}{[2\ln(b/a) - 1]},$$

And

$$\tau_b^{\min} \approx \frac{a}{u} \sqrt{\ln(b/a) - \frac{1}{2}}.$$

When $u \to \infty$, these relationships lead to the above optimal ratio of two stages. It is worth pointing out that the above result is only valid for 2D cases, and there is no general results for higher dimensions, except in some special 3D cases (Bénichou et al., 2006). Now let us use this limited results to help choose the possible values of algorithm-dependent parameters in FA (Yang, 2008, 2009), as an example.

For higher-dimensional problems, no result exists. One possible extension is to use extrapolation to get an estimate. Based on the results on 2D and 3D cases (Bénichou et al., 2011), we can estimate that for any d-dimensional cases $d \geq 3$

$$\frac{\tau_1}{\tau_2^2} \sim O\left(\frac{D}{a^2}\right), \quad \tau_m \sim O\left(\frac{b}{u}\left(\frac{b}{a}\right)^{d-1}\right),$$

where τ_m the mean search time or average number of iterations. This extension may not be good news for higher dimensional problems, as the mean number of function evaluations to find optimal solutions can increase exponentially as the dimensions increase. However, in practice, we do not need to find the guaranteed global optimality, we may be satisfied with suboptimality, and sometimes we may be 'lucky' to find such global optimality even with a limited/fixed number of iterations. This may indicate there is a huge gap between theoretical understanding and the observations as well as run-time behaviour in practice. More studies are highly needed to address these important issues.

5 NUMERICAL EXPERIMENTS

5.1 Landscape-based optimality: A 2D example

If we use the 2D simple, isotropic random walks for local exploration to demonstrate landscape-based optimality, then we have

$$D \approx \frac{s^2}{2},$$

where *s* is the step length with a jump during a unit time interval or each iteration step. From equation (9), the optimal ratio of exploitation and exploration in a special case of $b \approx 10a$ becomes

$$\frac{\tau_a}{\tau_b^2} \approx 0.2.$$

In case of $b/a \to \infty$, we have $\tau_a/\tau_b^2 \approx 1/8.$ which implies that more times should spend on the exploration stage. It is worth pointing out that the naive guess of 50–50 probability in each stage is not the best choice. More efforts should focus on the exploration so that the best solutions found by the algorithm can be globally optimal with possibly the least computing effort. However, this case may be implicitly linked to the implicit assumptions that the optimal solutions or search targets are multimodal. Obviously, for a unimodal problem, once we know its modality, we should focus more on the exploitation to get quick convergence.

In the case studies to be described below, we have used the FA to find the optimal solutions to the benchmarks. If set $\tau_b = 1$ as the reference timescale, then we found that the optimal ratio is between 0.15 to 0.24, which are roughly close to the above theoretical result.

5.2 Standing-wave function

Let us first use a multimodal test function to see how to find the fine balance between exploration and exploitation in an algorithm for a given task. A standing-wave test function can be a good example (Yang, 2009, 2010a)

$$f(x) = 1 + \left\{ \exp\left[-\sum_{i=1}^{d} \left(\frac{x_i}{\beta} \right)^{10} \right] - 2\exp\left[-\sum_{i=1}^{d} (x_i - \pi)^2 \right] \right\} \cdot \prod_{i=1}^{d} \cos^2 x_i,$$

which is multimodal with many local peaks and valleys. It has a unique global minimum at $f_{min} = 0$ at $(\pi, \pi, ..., \pi)$ in the domain $-20 \le x_i \le 20$ where $i = 1, 2, ..., d$ and $\beta = 15$. In this case, we can estimate that $R = 20$ and $a \approx \pi/2$, this means that $R/a \approx 12.7$, and we have in the case of $d = 2$

$$p_e \approx \tau_{\text{optimal}} \approx \frac{1}{2[2 - 1/\ln(R/a)]^2} \approx 0.19.$$

This indicate that the algorithm should spend 80% of its computational effort on global explorative search, and 20% of its effort on local intensive search.

For the FA, we have used $n = 15$ and 1,000 iterations. We have calculated the fraction of iterations/function evaluations for exploitation to exploration. That is, Q = exploitation/exploration, thus Q may affect the quality of solutions. A set of 25 numerical experiments have been carried out for each value of Q and the results are summarised in Table 1.

Table 1 Variations of Q and its effect on the solution quality

Q	0.4	0.3	0.2	0.1	0.05
f_{min}	9.4e-11	1.2e-12	2.9e-14	8.1e-12	9.2e-11

This table clearly shows that $Q \approx 0{:}2$ provides the optimal balance of local exploitation and global exploration, which is consistent with the theoretical estimation.

Though there is no direct analytical results for higher dimensions, we can expect that more emphasis on global exploration is also true for higher dimensional optimisation problems. Let us study this test function for various higher dimensions.

5.3 Comparison for higher dimensions

As the dimensions increase, we usually expect the number of iterations of finding the global optimality should increase. In terms of mean search time/iterations, Bénichou et al.'s (2006, 2011) intermittent search theory suggests that

$$\tau_m\Big|_{(d=1)} = \frac{2b}{u}\sqrt{\frac{b}{3a}},$$

$$\tau_m\Big|_{(d=2)} = \frac{2b^2}{au}\sqrt{\ln(b/a)},$$

$$\tau_m\Big|_{(d=3)} = \frac{2.2b}{u}\left(\frac{b}{a}\right)^2.$$

For higher dimensions, we can only estimate the main trend based on the intermittent search strategy. That is,

$$\frac{\tau_1}{\tau_2^2} \sim O\left(\frac{D}{a^2}\right), \quad \tau_m \sim O\left(\frac{b}{u}\left(\frac{b}{a}\right)^{d-1}\right),$$

which means that number of iterations may increase exponentially with the dimension *d*. It is worth pointing out that the optimal ratio between the two stage should be independent of the dimensions. In other words, once we find the optimal balance between exploration and exploitation, we can use the algorithm for any high dimensions.

Now let us use FA to carry out search in higher dimensions for the above standing wave function and compare its performance with the implication of intermittent search strategy. For the case of $b = 20$, $a = \pi/2$ and $u = 1$, Figure 3 shows the comparison of the numbers of iterations suggested by intermittent search strategy and the actual numbers of iterations using FA to obtain the globally optimal solution with a tolerance or accuracy of five decimal places.

It can be seen clearly that the number of iterations needed by the intermittent search strategy increases exponentially versus the number of dimensions, while the actual number of iterations used in the algorithm only increases slightly, seemingly weakly a low-order polynomial. This suggests that FA is very efficient and requires far fewer (and often many orders lower) number of function evaluations.

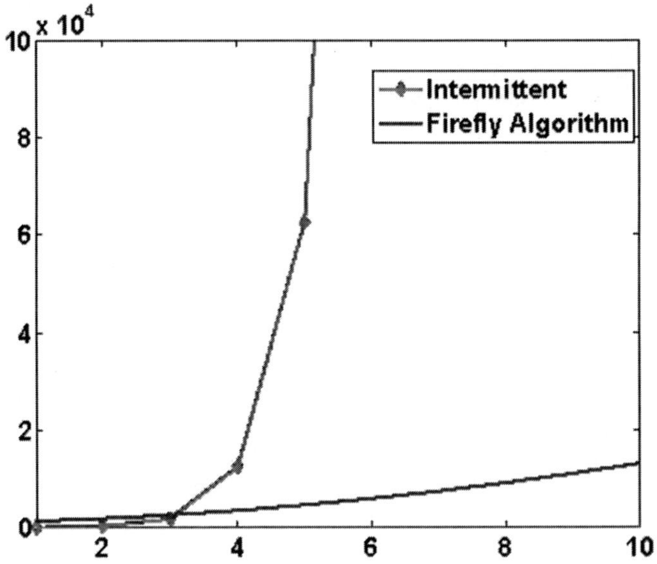

Note: This clearly show that FA is better than the intermittent search strategy.

Figure 3 Comparison of the actual number of iterations with the theoretical results by the intermittent search strategy (see online version for colours)

6 CONCLUSIONS

Nature-inspired metaheuristic algorithms have gained popularity, which is partly due to their ability of dealing with non-linear global optimisation problems. We have reviewed the fundamentals of FA, the latest developments with diverse applications. As the time of writing, a quick Google search suggests that there are about 323 papers on FAs from 2008. This review can only cover a fraction of the literature. There is no doubt that FA will be applied in solving more challenging problems in the near future, and its literature will continue to expand.

On the other hand, we have also highlighted the importance of exploitation and exploration and their effect on the efficiency of an algorithm. Then, we use the intermittent search strategy theory as a preliminary basis for analysing these key components and ways to find the possibly optimal settings for algorithm-dependent parameters.

With such insight, we have used the FA to find this optimal balance, and confirmed that FA can indeed provide a good balance of exploitation and exploration. We have also shown that FA requires far fewer function evaluations. However, the huge differences between intermittent search theory and the behaviour of metaheuristics in practice also suggest there is still a huge gap between our understanding of algorithms and the actual behaviour of metaheuristics. More studies in metaheuristics are highly needed.

It is worth pointing out that there are two types of optimality here. One optimality concerns that for a given algorithm what best types of problems it can solve. This is relatively easy to answer because in principle we can test an algorithm by a wide range of problems and then select the best type of the problems the algorithm of interest can solve. On other hand, the other optimality concerns that for a given problem what best algorithm is to find the solutions efficiently. In principle, we can compare a set of algorithms to solve the same optimisation problem and hope to find the best algorithm(s). In reality, there may be no such algorithm at all, and all test algorithms may not perform well. Search for new algorithms may take substantial research efforts.

The theoretical understanding of metaheuristics is still lacking behind. In fact, there is a huge gap between theory and applications. Though theory lags behind, applications in contrast are very diverse and active with thousands of papers appearing each year. Furthermore, there is another huge gap between small-scale problems and large-scale problems. As most published studies have focused on small, toy problems, there is no guarantee that the methodology that works well for such toy problems will work for large-scale problems. All these issues still remain unresolved both in theory and in practice.

As further research topics, most metaheuristic algorithms require good modifications so as to solve combinatorial optimisation properly. Though with great interest and many extensive studies, more studies are highly needed in the area of combinatorial optimisation using metaheuristic algorithms. In addition, most current metaheuristic research has focused on small scale problems, it will be extremely useful if further research can focus on large-scale real-world applications.

REFERENCES

1. Abshouri, A.A. , Meybodi, M.R. , Bakhtiary, A. (2011). 'New firefly algorithm based on multiswarm and learning automata in dynamic environments'. Third Int. Conference on Signal Processing Systems (ICSPS2011), August 27–28, Yantai, China, 73-77

2. Apostolopoulos, T. , Vlachos, A. (2011). 'Application of the firefly algorithm for solving the economic emissions load dispatch problem'. International Journal of Combinatorics. Article ID 523806, [online] http://www.hindawi.com/journals/ijct/2011/523806.html

3. Azad, S.K. , Azad, S.K. (2011). 'Optimum design of structures using an improved firefly algorithm'.International Journal of Optimisation in Civil Engineering. 1, 2, 327-340

4. Banati, H. , Bajaj, M. (2011). 'Firefly based feature selection approach'. Int. J. Computer Science Issues. 8, 2, 473-480

5. Bansal, J.C. , Deep, K. (2008). 'Optimisation of directional overcurrent relay times by particle swarm optimisation'. in Swarm Intelligence Symposium (SIS 2008), IEEE Publication, 1-7

6. Basu, B. , Mahanti, G.K. (2011). 'Firefly and artificial bees colony algorithm for synthesis of scanned and broadside linear array antenna'. Progress in Electromagnetic Research B. 32, 2, 169-190

7. Bénichou, O. , Loverdo, C. , Moreau, M. , Voituriez, R. (2006). 'Two-dimensional intermittent search processes: an alternative to Lévy flight strategies'. Phys. Rev.. E74, 020102(R)

8. Bénichou, O. , Loverdo, C. , Moreau, M. , Voituriez, R. (2011). 'Intermittent search strategies'. Review of Modern Physics. 83, 1, 81-129

9. Blum, C. , Roli, A. (2003). 'Metaheuristics in combinatorial optimisation: overview and conceptual comparison'. ACM Comput. Surv.. 35, 3, 268-308

10. Chatterjee, A. , Mahanti, G.K. , Chatterjee, A. (2012). 'Design of a fully digital controlled reconfigurable switched beam concentric ring array antenna using firefly and particle swarm optimisation algorithm'.Progress in Electromagnetic Research B. 36, 1, 113-131

11. Coelho, L.d.S. , Bernert, D.L.d.A. , Mariani, V.C. (2011). 'A chaotic firefly algorithm applied to reliability-redundancy optimisation'. in 2011 IEEE Congress on Evolutionary Computation (CEC '11), 517-521

12. Durkota, K. (2011). Implementation of a Discrete Firefly Algorithm for the QAP Problem within the Sage Framework. Czech Technical University, BSc thesis

13. Farahani, S.M. , Abshouri, A.A. , Nasiri, B. , Meybodi, M.R. (2011a). 'A Gaussian firefly algorithm'. Int. J. Machine Learning and Computing. 1, 5, 448-453

14. Farahani, S.M. , Nasiri, B. , Meybodi, M.R. (2011b). 'A multiswarm based firefly algorithm in dynamic environments'. in Third Int. Conference on Signal Processing Systems (ICSPS2011), August 27–28, Yantai, China, 68-72

15. Fister, I., Jr. , Fister, I. , Brest, J. , Yang, X.S. , Filipič, B. , Šilc, J. (2012). 'Memetic firefly algorithm for combinatorial optimisation'. Bioinspired Optimisation Methods and Their Applications (BIOMA2012), 24–25 May 2012, Bohinj, Slovenia, 75-86

16. Floudas, C.A. , Pardolos, P.M. (2009). Encyclopedia of Optimisation. 2nd ed., Berlin, Germany:Springer

17. Gandomi, A.H. , Yang, X.S. , Alavi, A.H. (2011). 'Cuckoo search algorithm: a metaheuristic approach to solve structural optimisation problems'. Engineering with Computers. 27, 10.1007/s00366-011-0241-y

18. Giannakouris, G. , Vassiliadis, V. , Dounias, G. (2010). 'Experimental study on a hybrid nature-inspired algorithm for financial portfolio optimisation'. SETN 2010, Lecture Notes in Artificial Intelligence (LNAI 6040),101-111

19. Hassanzadeh, T. , Vojodi, H. , Moghadam, A.M.E. (2011). 'An image segmentation approach based on maximum variance intra-cluster method and firefly algorithm'. in Proc. of 7th Int. Conf. on Natural Computation (ICNC2011). 1817-1821

20. Horng, M-H. , Lee, Y-X. , Lee, M-C. , Liou, R-J. , Parpinelli, R. , Lopes, H.S. (2012). 'Firefly metaheuristic algorithm for training the radial basis function network for data classification and disease diagnosis'. Theory and New Applications of Swarm Intelligence. 115-132

21. Horng, M-H. (2012). 'Vector quantization using the firefly algorithm for image compression'. Expert Systems with Applications. 39, 1, 1078-1091

22. Horng, M-H. , Liou, R-J. (2011). 'Multilevel minimum cross entropy threshold selection based on the firefly algorithm'. Expert Systems with Applications. 38, 12, 14805-14811

23. Jati, G.K. , Suyanto, S. (2011). 'Evolutionary discrete firefly algorithm for travelling salesman problem'.ICAIS2011, Lecture Notes in Artificial Intelligence (LNAI 6943). 393-403

24. Kennedy, J. , Eberhart, R. (1995). 'Particle swarm optimisation'. in Proc. of the IEEE Int. Conf. on Neural Networks. Piscataway, NJ, 1942-1948

25. Nandy, S. , Sarkar, P.P. , Das, A. (2012). 'Analysis of nature-inspired firefly algorithm based back-propagation neural network training'. Int. J. Computer Applications. 43, 22, 8-16

26. Palit, S. , Sinha, S. , Molla, M. , Khanra, A. , Kule, M. (2011). 'A cryptanalytic attack on the knapsack cryptosystem using binary firefly algorithm'. in 2nd Int. Conference on Computer and Communication Technology (ICCCT), 15–17 September, India, 428-432

27. Parpinelli, R.S. , Lopes, H.S. (2011). 'New inspirations in swarm intelligence: a survey'. Int. J. Bio-Inspired Computation. 3, 1, 1-16 Abstract

28. Rajini, A. , David, V.K. (2012). 'A hybrid metaheuristic algorithm for classification using micro array data'. Int. J. Scientific & Engineering Research. 3, 2, 1-9

29. Rampriya, B. , Mahadevan, K. , Kannan, S. (2010). 'Unit commitment in deregulated power system using Lagrangian firefly algorithm'. Proc. of IEEE Int. Conf. on Communication Control and Computing Technologies (ICCCCT2010). 389-393

30. Sayadi, M.K. , Ramezanian, R. , Ghaffari-Nasab, N. (2010). 'A discrete firefly meta-heuristic with local search for makespan minimization in permutation flow shop scheduling problems'. Int. J. of Industrial Engineering Computations. 1, 1, 1-10

31. Senthilnath, J. , Omkar, S.N. , Mani, V. (2011). 'Clustering using firefly algorithm: performance study'. Swarm and Evolutionary Computation. 1, 3, 164-171

32. Yang, X.S. (2008). Nature-Inspired Metaheuristic Algorithms. UK:Luniver Press

33. Yang, X.S. , Watanabe, O. , Zeugman, T. (2009). 'Firefly algorithms for multimodal optimisation'. Proc. 5th Symposium on Stochastic Algorithms, Foundations and Applications, Lecture Notes in Computer Science. 5792, 169-178

34. Yang, X.S. (2010a). Engineering Optimisation: An Introduction with Metaheuristic Applications. USA:John Wiley and Sons

35. Yang, X.S. , Gonzalez, J.R. , et al. (2010b). 'A new metaheuristic bat-inspired algorithm'. 284, Nature Inspired Cooperative Strategies for Optimisation (NICSO 2010), Springer, SCI, 65-74

36. Yang, X.S. (2010c). 'Firefly algorithm, stochastic test functions and design optimisation'. Int. J. Bio-Inspired Computation. 2, 2, 78-84 Abstract

37. Yang, X.S. , Deb, S. (2009). 'Cuckoo search via Lévy flights'. Proceedings of World Congress on Nature & Biologically Inspired Computing (NaBIC 2009, India). USA:IEEE Publications , 210-214

38. Yang, X.S. (2011). 'Chaos-enhanced firefly algorithm with automatic parameter tuning'. Int. J. Swarm Intelligence Research. 2, 4, 1-11

39. Yang, X.S. , Parpinelli, R. , Lopes, H.S. (2012a). 'Swarm-based metaheuristic algorithms and no-free-lunch theorems'. Theory and New Applications of Swarm Intelligence. Intech Open Science, 1-16

40. Yang, X.S. (2012b). 'Multiobjective firefly algorithm for continuous optimization'. Engineering with Computers. Online Fist,

41. Yang, X.S. , Deb, S. , Fong, S. (2011). 'Accelerated particle swarm optimization and support vector machine for business optimization and applications'. 136, Networked Digital Technologies (NDT 2011), Communications in Computer and Information Science, Part I, 53-66

42. Yousif, A. , Abdullah, A.H. , Nor, S.M. , abdelaziz, A.A. (2011). 'Scheduling jobs on grid computing using firefly algorithm'. J. Theoretical and Applied Information Technology. 33, 2, 155-164

43. Zaman, M.A. , Matin, M.A. (2012). 'Nonuniformly spaced linear antenna array design using firefly algorithm'.Int. J. Microwave Science and Technology. 2012, Article ID: 256759, 10.1155/2012/256759 8 pp

Chapter 7

A Localization Method for Underwater Wireless Sensor Networks Based on Mobility Prediction and Particle Swarm Optimization Algorithms

Ying Zhang [1], Jixing Liang [1], Shengming Jiang [1] and Wei Chen [2,*]

[1]College of Information Engineering, Shanghai Maritime University, Shanghai 201306, China

[2]Department of Computer Science, Tennessee State University, Nashville, TN 37209, USA

ABSTRACT

Due to their special environment, Underwater Wireless Sensor Networks (UWSNs) are usually deployed over a large sea area and the nodes are usually floating. This results in a lower beacon node distribution density, a longer time for localization, and more energy consumption. Currently most of the localization algorithms in this field do not pay enough consideration on the mobility of the nodes. In this paper, by analyzing the mobility patterns of water near the seashore, a localization method for UWSNs based on a Mobility Prediction and a Particle Swarm Optimization algorithm (MP-PSO) is proposed. In this method, the range-based PSO algorithm is used to locate the beacon nodes, and their velocities can be calculated. The velocity of an unknown node is calculated by using the spatial correlation of underwater object's mobility, and then their locations can be predicted. The range-based PSO algorithm may cause considerable energy consumption and its computation complexity is a little bit high, nevertheless the number of beacon nodes is relatively smaller, so the calculation for the large number of unknown nodes is succinct, and this method can obviously decrease the energy consumption and time cost of localizing these mobile nodes. The simulation results indicate that this method has higher localization accuracy and better localization coverage rate compared with some other widely used localization methods in this field.

KEYWORDS

underwater wireless sensor networks; mobility patterns; localization; mobility prediction; particle swarm optimization

1. INTRODUCTION

The marine economy has witnessed rapid development recently, and marine rights and interests are receiving more and more attention in many countries, therefore research on UWSNs has developed quickly, providing basic technical support to many application fields, such as ocean environment monitoring, resource exploration, natural disaster warning and military defense [1]. Besides the basic characteristics of ordinary WSNs, like large-scale deployment and limited energy, underwater WSNs have some differences with terrestrial ones. First, underwater communication can only be realized by acoustic signals, which have a lower bandwidth and higher error rates. Second, the scale of node deployment in underwater environments is larger and the beacon nodes are sparser. In addition, because of water currents, there is the non-negligible node mobility which may cause frequent changes to the network topology [2,3].

The localization of mobile nodes is important and necessary for underwater sensor networks. In the marine military defense area, the identification and tracking of the intrusion objects must rely on the nodes' location. It can also be used for animal tracking in the underwater environment, where the localization of mobile nodes can help researchers to determine marine animals' locations and movements easily, and in particular mobile nodes can extend the tracking area and better realize real-time tracking. In marine environmental monitoring, the detection area is usually very large and dynamic. With random mobility, the sensor nodes could realize the monitoring of the whole area. In this process, the localization of mobile nodes is necessary for identifying where the polluted areas will be. In addition, the path-planning and the coverage optimization of the underwater nodes cannot be realized without the localization of the mobile nodes [4,5]. The information collection and transmission are realized by the sensor nodes, and the relevant data without location will make no sense, so the localization of mobile sensor nodes has gradually become a research hotspot and focus.

However, the characteristics of UWSN described above present great difficulties and challenges to the localization of underwater mobile nodes in a large-scale network. Acoustic communication has a bigger propagation delay, lower bandwidth and higher error rate compared to radio communication. This imposes more limits on any localization algorithm. The algorithms based on transmission of tremendous amounts of data or higher real-time communication requirements will not be applicable in this kind of circumstance. In addition, the mobility of nodes makes the topology change frequently. The algorithms designed for static networks need to run the localization procedure periodically to update the nodes' location, which may cause more energy consumption for communication in an underwater environment. Moreover, the batteries of underwater sensor nodes can almost never be replaced and the energy is strictly limited, so under normal circumstances, it is difficult to achieve higher localization accuracy and localization coverage rate in an underwater environment [6,7].

As shown in Figure 1, surface buoys equipped with GPS can obtain their locations. The number of beacon nodes is much less than the number of unknown nodes. Beacon nodes have more energy and their communication radius is about 200 m, so they can communicate directly with the buoys and have more neighbor nodes. In addition, beacon nodes have more hardware resources and better computing ability compared to unknown nodes, so they can use range-based localization methods. The unknown node are relatively cheaper and one does not want them to waste their energy. The communication radius of an unknown node is about 100 m and it cannot communicate with the buoys directly. Typically, it can only connect to its local (usually one-hop) neighbors. With local information exchange among themselves and nearby beacon nodes, the unknown nodes can achieve local positioning so as to effectively participate in the network activities.

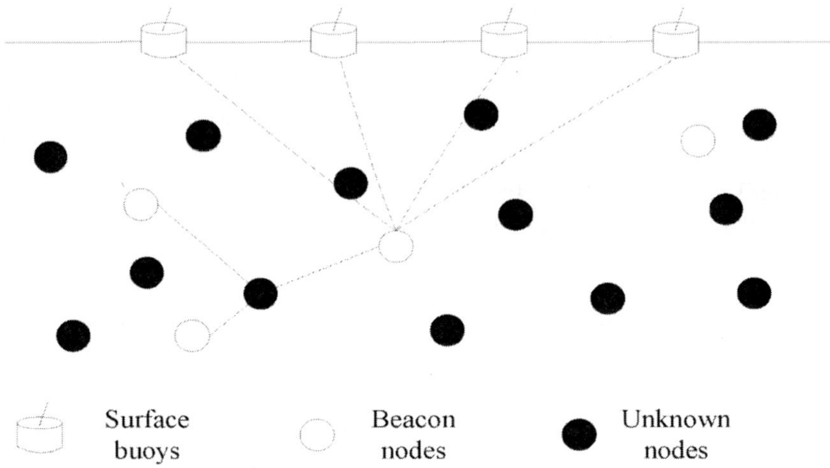

Figure 1. Schematic diagram of the monitoring model of an UWSN.

Aiming at the special environment of UWSN and the difficulties in mobile node localization, we propose a method based on the mobility prediction and PSO algorithms. First we measure the distances between the beacon nodes and buoys, and locate the beacon nodes by using a range-based PSO algorithm which has a relative high accuracy. According to the location results at two time points, we can calculate the velocity of beacon nodes in the last instant. The velocities of the unknown nodes can be estimated by the velocity information of beacon nodes within the communication scope and the spatial correlation of underwater objects, and then the location of the unknown nodes for the next moment can be predicted by the mobility prediction method. Considering the higher energy consumption a the range-based algorithm, we just use it to locate the beacon nodes, whose number is far fewer compared to the unknown nodes, and on this basis, the localization of the unknown nodes can achieve a relatively high-accuracy result with fine instantaneity by the mobility prediction method. The rest of this article is organized as follows: in Section 2, some background knowledge and related work on nodes localization are introduced. Then, in Section 3, we describe the MP-PSO method in detail. Section 4 is the simulation and results analysis, and finally we conclude the paper in Section 5.

2. RELATED WORKS

Localization methods for wireless sensor networks can be divided into two types: range-based and range-free methods. The range-based methods such as the received signal strength indicator (RSSI), time difference of arrival (TDOA) and time of arrival (TOA) use hardware to measure the distance information. These kinds of method have a higher accuracy, but they increase the network cost and energy consumption [8]. The range-free methods use the connectivity of the network to locate the unknown nodes. The typical range-free methods mainly include the DV-HOP, Convex Programming and Centroid Localization algorithm. These methods have no additional hardware requirements, and they have lower energy consumption and shorter positioning time, but their accuracy is usually lower [9].

There are many researches on terrestrial nodes localization. In [10] the authors proposed a range-free localization algorithm based on a sequential Monte Carlo localization method. It can exploit mobility to improve the localization accuracy. In [11], a Monte Carlo localization algorithm with mobility prediction (MCL-MP) was proposed, and it can further improve the accuracy by using prediction and filtering for the unknown nodes based on dynamic sampling.

Studies on the localization of underwater mobile nodes always face some challenges, and most of them were designed for small-scale networks. For example, GPS Intelligent Buoys (GIB) based on surface buoys and one-hop communication under the water have been proposed. This approach has a high accuracy but the hardware is complex and the cost is high [12]. In [13], a so-called Silent Localization algorithm was proposed, which does not need time synchronization and is applicable to one-hop underwater networks. In addition, in [14], a Scalable Localization with Mobility Prediction (SLMP) method was proposed, and it is closer to localization in an actual environment. The SLMP algorithm has two stages to locate the unknown nodes. First, the velocity of beacon nodes will be estimated by the Durbin algorithm to perform the online linear prediction. The unknown nodes are then located by using the mobility prediction based on the spatial correlation of sensor nodes.

A three-dimensional deployment space is another important characteristic of UWSNs. In [15], an efficient localization scheme which can transform the three-dimensional localization problem into a two-dimensional counterpart via a projection technique was proposed; it can make all the nodes map to the same plane. Because the depth information can be obtained by a pressure sensor, this scheme not only makes the two-dimensional localization algorithm apply in the three-dimensional space, but also simplifies the amount of calculation for three-dimensional localization, and reduces the energy consumption of the process. In this paper, we also use this method to project the nodes on a certain plane, and only consider the velocity of the nodes in two orthotropic directions. The principle of this method is shown in Figure 2.

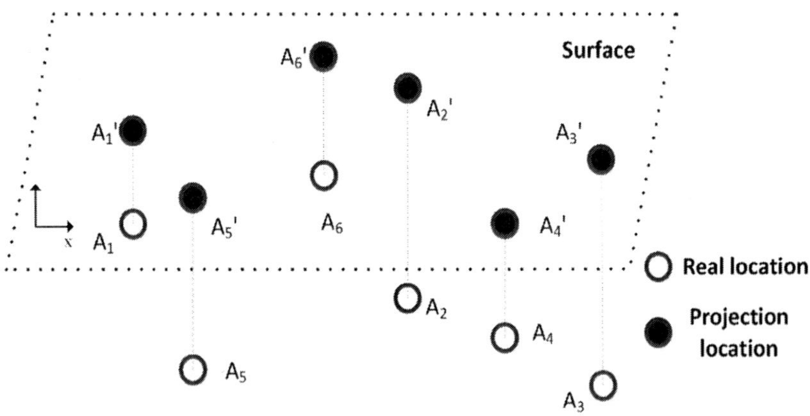

Figure 2. A schematic diagram of the projection method.

As shown in Figure 2, A_1–A_6 are sensor nodes in an underwater networks. We project them to the water surface and the corresponding locations are A_1'–A_6'. Then we only need to locate the nodes in the surface plane, and the final location can be obtained by simply adding the respective depth information.

3. MP-PSO LOCALIZATION METHOD

The proposed MP-PSO method is a type of multi-step localization algorithm, and it can be divided into the beacon node localization and the unknown node localization. The beacon nodes are located by measuring the distances from the nodes to the buoys on the water surface with the range-based PSO algorithm, and it can achieve a higher localization accuracy. The locations of unknown nodes in the next moment can be predicted by estimating the speed of movement of the unknown nodes.

3.1. The Node Mobility Model

Research shows that the mobility of underwater object is influenced by water current, temperature and some other factors [16], so we cannot use a unified model to describe the nodes mobility for all environments. However, under a specific environment, the nodes mobility is not completely random, so we can establish a specific mobility model for this specific environment according to the temporal and spatial correlation. Nowadays the UWSN is mainly used in seashore environment, where the water is relatively shallow, usually it is less than 100 m, and the water current is relatively flat, so the nodes' mobility situation is not complex. In addition, the underwater nodes' mobility in seashore area has an obvious characteristic, namely it demonstrates a certain semi periodic property because of tides [17].

In [14], a mobility model based on the Euler algorithm was proposed, which takes the seashore environment as the application background. It assumes the flow velocity field is superposed by the tidal field and the remnant flow field. The tidal field is assumed to be oscillating homogeneously in one direction, and the remnant flow field is assumed to be an infinite sequence alternately rotating clockwise and anticlockwise by turns. The dimensionless velocity field in the kinematical model can be approximated as:

$$Vx(t) = k_1 \cdot \lambda \cdot v \cdot \sin(k_2 \cdot x) \cdot \cos(k_3 \cdot y) + k_1 \cdot \lambda \cdot \cos(2k_1 \cdot t) + k_4 \quad (1)$$

$$Vy(t) = -\lambda \cdot v \cdot \cos(k_2 \cdot x) \cdot \sin(k_3 \cdot y) + k_5 \quad (2)$$

where *Vx* is the speed of the *X* axis direction and *Vy* is the speed of the *Y* axis direction, λ is the ratio of tidal excursion and the diameter of the remnant vortex. Here, λ, k_1, k_2, k_3 and v are related to factors like tides, temperature and salinity, and k_5 and k_4 are random variables. The values of these parameters are set in the simulation part of this paper. Figure 3shows the velocity variation of an underwater object with time in a typical seashore environment.

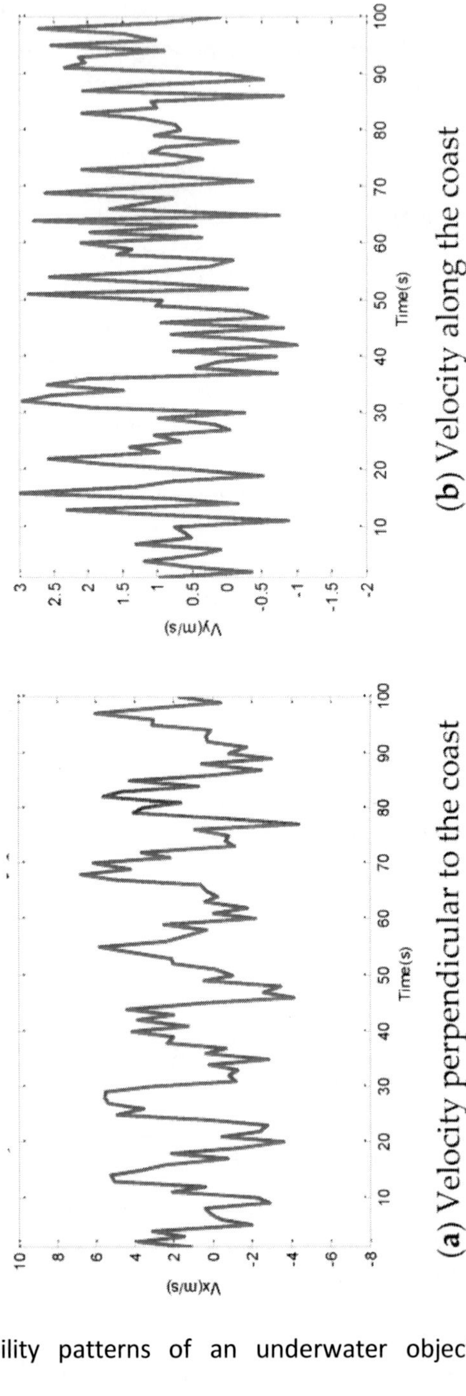

Figure 3. Mobility patterns of an underwater object in a seashore environment.

This model is designed for the seashore environment, so it can evaluate the mobility of nodes whose velocity is relatively constant. As shown in Figure 3, the velocity in the two directions has an obvious semiperiodic property and temporal correlation, which can guarantee the practicability of mobility prediction of the unknown nodes' localization in the water environment.

3.2. Propagation Loss Model for Submarine Sound Signal

In an underwater acoustic communication channel, the strength of a signal received by the unknown nodes: S_r can be calculated as Equation (3). S_t is the strength of the signal transmitted by beacon nodes. *A*(*d*,*f*) is the total attenuation for the acoustic signal in the transmission procedure:

$$S_r = S_t - A(d,f) \tag{3}$$

Influenced by the environmental factors, the transmission attenuation *A*(*d*,*f*) is the sum of the extended losses and the absorption losses:

$$A(d,f) = d^{\delta} \cdot \alpha^{d};\ \ \alpha = 10^{\alpha(f)/10} \tag{4}$$

where, *d* is the transmission distance, *α* is the absorption coefficient, and δ is the attenuation coefficient (generally valued as 1 < δ < 2). *α*(*f*) is absorption loss, it can be approximately estimated based on Thorp algorithm [18], and it can be calculated as Equation (5):

$$\alpha(f) = 0.11\frac{f^2}{1+f^2} + 44\frac{f^2}{4100+f^2} + 2.75\times10^{-4}f^2 + 0.003 \tag{5}$$

where, *f* is the frequency of transmission signal of the nodes. Referring to the frequency of underwater network nodes widely used currently, the frequency can be set as 50 kHz, and the transmission power can be set as 2 W, namely 33 dBm [19].

3.3. Localization of Beacon Nodes

The swarm search strategy has the feature of adaptive optimization and the PSO algorithm is an efficient method which can realize accurate range-based localization for WSNs. The authors of [20] proposed a distributed node localization method on the basis of a multivariable optimization algorithm based on biogeography and the Particle Swarm Optimization algorithm, which can effectively improve the localization accuracy of the nodes. In [21], a binary Particle Swarm Optimization algorithm for distributed node localization in WSNs was proposed. Each unknown node performs localization by the measurement of distances from three or more neighbor beacon nodes to the unknown node. The nodes which already determined their locations will be regarded as the reference nodes for the residual nodes, and they will be involved in the subsequent localization procedure.

Considering the defect that the traditional PSO algorithm can easily fall into a local minimum and has lower convergence speed [22], in this paper, an inertia weight with Gaussian function decline and competition mechanism are adopted to improve the search ability, convergence speed and execution efficiency of the PSO. The improved PSO algorithm can realize the accurate localization of beacon nodes by the following steps:

Step 1: Initialize the population. The number of particles is *Pnum*. Calculate the fitness of each particle in Equation (6). The individual extremum is denoted as P_i^t, and the global extremum is denoted as P_g^t.

$$f(t) = \frac{1}{H}\sum_{i=1}^{H}\left|\sqrt{(x-x_i)^2+(y-y_i)^2}-l_i\right| \tag{6}$$

where, H is the number of effective buoys, (x, y) is the two-dimensional coordinate of beacon node, (x_i, y_i) is the coordinate of buoy, and l_i is the distance from this beacon node to every buoy.

Step 2: Calculate the inertia weight with Equation (7), and update the position and velocity with Equations (8) and (9).

$$w(t) = (w_{\max} - w_{\min})\exp[-\frac{t^2}{(k+t_{\max})^2} + w_{\min}] \tag{7}$$

$$V_i^{t+1} = V_i^t + c_1 r_1 (P_i^t - X_i^t) + c_2 r_2 (P_g^t - X_i^t) \quad (8)$$

$$X_i^{t+1} = X_i^t + V_i^{t+1} \quad (9)$$

where, *t* is the iterations, c_1 and c_2 are the learning factors. r_1 and r_2 are the random numbers uniformly distributed in (0,1). c1r1(Pti−Xti) and c2r2(Ptg−Xti) are the evolutions of a particle towards the local optimum and global optimum, respectively, and they can be regarded as the variation of velocity updated by theparticle in this iteration.

Step 3: Calculate the fitness of each particle and get the average fitness, then update the individual extremum Pti and the global extremum Ptg. Eliminate the particles whose fitness is larger than double average fitness.

Step 4: Judge whether it satisfies the condition: $P_g^t < \varepsilon$ (ε is the threshold of position error) or $t = t_{max}$. If one of the condition is satisfied, turn to Step 5, or $t = t + 1$, and turn to Step 2.

Step 5: Output the global optimal solution, and get the coordinates of the beacon nodes.

3.4. Localization of Unknown Nodes

Because the communication and computation ability of the underwater unknown nodes is limited, it is too difficult for them to perform the range-based localization algorithm. In this paper, by the proposed MP-PSO method, we use the temporal and spatial correlation of the underwater objects to realize the unknown nodes' mobility prediction.

3.4.1. The Calculation of Beacon Nodes' Velocity

The range-based localization algorithm needs more energy and computation. In real practical applications, the number of beacon nodes in real networks is relatively less. These nodes have rather strong computational ability and more energy, so they can run the localization algorithm periodically to get their velocities of the previous time points.

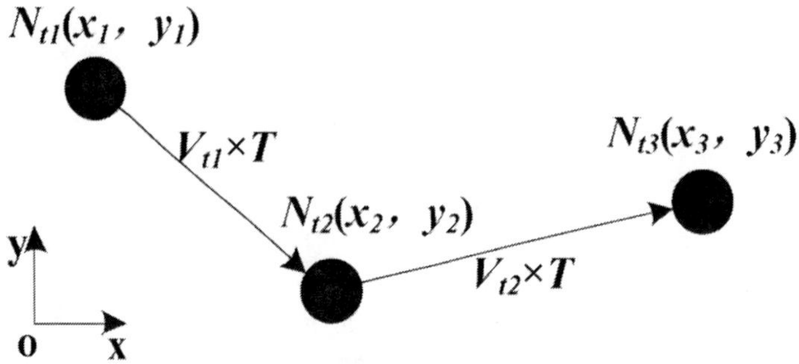

Figure 4. The calculation of beacon nodes' velocity.

As shown in Figure 4, the beacon node N moves from N_{t1} to N_{t2} and then to N_{t3}, and the position offsets of the two movements are $V_{t1} \times T$ and $V_{t2} \times T$ respectively. The coordinates of N at t_1 and t_2 can be obtained by the PSO algorithm, and then the velocity at t_1 for the beacon node can be calculated by Equation (10):

$$v_x = \frac{x_2 - x_1}{T}; \ v_y = \frac{y_2 - y_1}{T} \tag{10}$$

where T is the localization period, and the value of T may affect the localization result. Namely, if the value of T is too large, the estimation of the velocity of the beacon node may be imprecise. The instantaneity of velocity for the underwater object is strong, so if we use the velocity at a certain time to represent the velocity of a period in the past, it may cause large errors. On the other hand, if the T value is too small, the localization algorithm running frequency will be too high, which may cause more energy consumption and computational overhead.

3.4.2. The Calculation of Unknown Nodes' Velocity

The beacon nodes broadcast a packet in the network after getting their velocities, and the packet contains the identity, velocity and time identification information. Time identification represents the moment of positioning for this node. Since all the nodes cannot complete their positioning at the same time, the referenced nodes selected should be in the same positioning round as the unknown nodes, that is to say they should have the same time identification. The unknown node *W*receives the packets from different beacon nodes, and it will sign the beacon node which has the same time identification as a referenced node. Finally the list of reference nodes will be set up, and it includes the identity, velocity, time identification information, and the received signal strength. The unknown node *W* can then calculate its mobile velocity with Equation (11):

$$\begin{cases} v_x(w) = \sum_{i=1}^{M} \varsigma_{iw} v_x(i) \\ v_y(w) = \sum_{i=1}^{M} \varsigma_{iw} v_y(i) \end{cases} \tag{11}$$

where, *M* is the number of referenced nodes, ς_{iw} is the weight of referenced nodes, and it can be calculated by Equation (12):

$$\varsigma_{iw} = \frac{r_{iw}}{\sum_{i=1}^{M} r_{iw}} \tag{12}$$

where, r_{iw} is the signal strength received from the referenced node *i* by unknown node *W*. This method can decrease the errors of the referenced nodes which are far away from node *W*, and make the velocity of *W* more close to the real value. Considering the beacon nodes are relatively sparse in the underwater networks, we adopt a cooperative mechanism to calculate the velocity of the unknown nodes. The unknown nodes which already get their velocities can be taken as the new reference nodes. The principle of this method is shown in Figure 5.

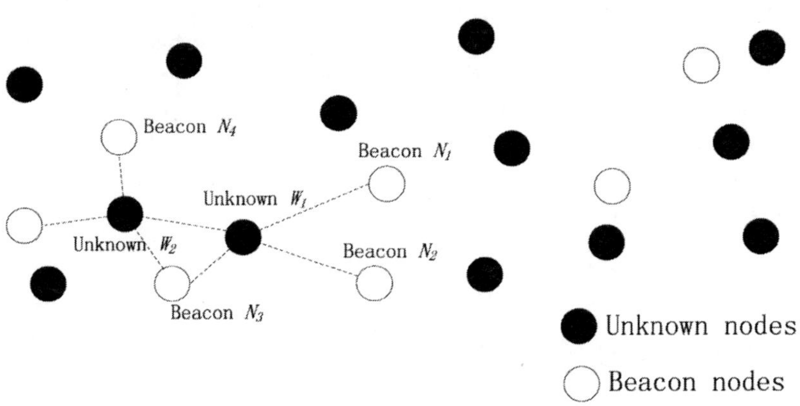

Figure 5. The calculation of unknown nodes' velocity.

As shown in Figure 5, the unknown node W_1 calculates its velocity using beacon nodes N_1, N_2 and N_3, and then W_1broadcasts a packet containing the identity, velocity and time identification information, so the unknown node W_2 can calculate its velocity by using beacon nodes N_3, N_4, N_5 and the unknown node W_1 as the reference nodes.

The velocities of unknown nodes have relatively more errors, especially when the new referenced nodes are far away from the unknown node, it will cause larger error accumulation. In order to further decrease the error, the unknown node will distribute a confidence coefficient to each referenced node. We sort the referenced nodes according to their confidence coefficient from the biggest to the smallest, and select the first *M* nodes of among them to participate in the velocity calculation. According to Equations (11) and (12), we can know that the number of referenced nodes can affect the velocity calculation. If the value of *M* is too small, the velocity of the unknown nodes will be calculated by only a few nodes, which cannot take full use of the spatial correlation of underwater objects' mobility. On the other hand, if the value of *M* is too large, the nodes which are far away from the unknown node will be involved in the velocity calculation, and it will cause large errors, so the value of *M* is determined by the number of all the nodes and the proportion of beacon nodes in the network. The confidence coefficient can be calculated by Equation (13):

$$\eta_{iw} = \frac{r_{iw}}{r_{w\max}} - 0.05k - 0.1r \tag{13}$$

where, r_{iw} is the signal strength received from reference node *i* by unknown node *W*, r_{wmax} is the maximum of r_{iw}, *k* is the number of unknown nodes which are involved in the calculation of the velocity for this reference node, and *r* is the times that the node is taken as the reference node. As an example in Figure 5, the values of *k* and *r* for beacon nodes are all 0. When we calculate the velocity of W_2, the values of *k* and *r* of reference node W_1 are 0 and 1, respectively. When the node W_2works as a reference node, the value of *k* is 1, and the value of *r* is 2.

3.4.3. Updating of Unknown Nodes' Location

After calculating the unknown nodes' velocity, we can update their locations. In fact, we just need to store the velocity of each node and update the location when it is necessary. The location can be updated with Equation (14):

$$\begin{cases} x\prime = x + \sum_{i=1}^{n} v_{xi} \cdot T \\ y\prime = y + \sum_{i=1}^{n} v_{yi} \cdot T \end{cases} \tag{14}$$

where, T is the positioning period, and ∑ni=1vi·t is the displacement between two positioning moments.

3.5. The Process of MP-PSO Method

As described above, the MP-PSO method is a kind of multi-steps localization algorithm, and the steps can be described as follows:

Step 1: Project the nodes into the plane which contains the surface buoys, and transform the three-dimensional localization problem into a two-dimensional localization.

Step 2: Locate the beacon nodes by using the range-based PSO algorithm.

Step 3: Calculate the velocity of beacon nodes according to the localization results at two instants.

Step 4: The unknown nodes sort their reference nodes in descending order according to the confidence value.

Step 5: If the number of reference nodes is more than *M*, then select the *M* reference nodes with larger confidence coefficients to calculate the velocity, if not, use all the reference nodes to calculate the velocity.

Step 6: Update the locations of the unknown nodes.

The MP-PSO algorithm process is shown in Figure 6.

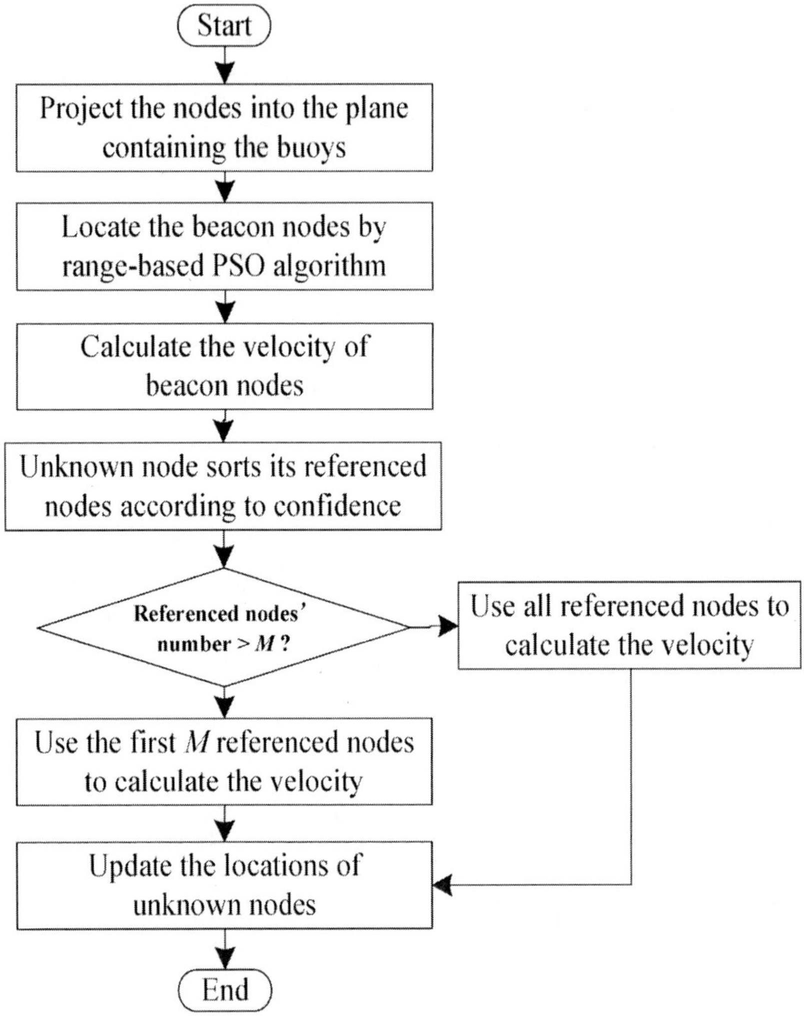

Figure 6. The process of MP-PSO algorithm.

4. SIMULATIONS AND ANALYSIS

4.1. Parameter Setting

We use MATLAB R2009a as the simulation platform to execute the algorithm in this paper. The settings of the simulation environment and related parameters are listed in Table 1.

Table 1. Simulation Parameters Setting.

Parameters	Values Setting
Deployment range	1000 m × 1000 m
Localization period	1 s
Communication radius of beacon nodes	200 m
Communication radius of unknown nodes	100 m
Transmission power	2 W (33 dBm)
Transmission frequency	50 kHz
Attenuation coefficient	2
Buoy number	20
Node number	200
v	$N(1,(0.1)^2)$
k_1,k_2	$N(\pi,(0.1\pi)^2)$
λ	$N(3,(0.3)^2)$
K_3	$N(2\pi,(0.2\pi)^2)$
K_4,k_5	$N(1,(0.1)^2)$

There are 20 surface buoys and 200 underwater nodes deployed in the monitoring area, the range of the area is 1000 m × 1000 m, and the depth of water is less than 100 m. We just investigate the moving state of the nodes driven by the water currents and waves in the seashore within 100 s after the initial deployment. In Table 1, the value of variables v, k_1–k_5 and λcomplies with the normal distribution, which can make the mobility of the nodes more close to the real environment.

According to the discussion in Section 3.4.2, the number of reference nodes can affect the localization result, so we need to select the best value of M by simulation testing. Keeping the parameter settings as in Table 1, by changing the proportion of beacon nodes to 10%, 15%, 20%, 25% and 30%, respectively, we can get the relationship of the average errors of the unknown nodes' velocities with different values of M after 100 s. The result is shown in Figure 7.

From Figure 7, it can be seen that the velocity error first decreases, and then it begins to increase as the number of reference nodes increases. This is accord with the above analysis of the value of M. When the value of M is too large, the reference nodes whose confidences are smaller will participate in the velocity calculation, and may cause a large error accumulation. Figure 7 also shows that the optimal value of the number of reference nodes is related to the proportion of beacon nodes in the network. If the beacon node proportion is different, then the optimal value of *M* is also different. In the subsequent simulations, we select the value of *M* according to Figure 7.

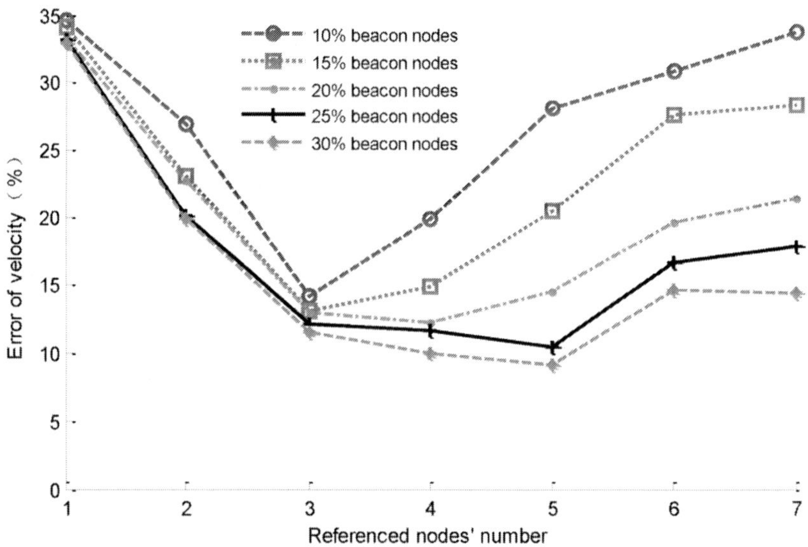

Figure 7. The relationship of velocity error and *M*.

4.2. Localization Results

Keeping the settings of Table 1, we set the ratio of beacon nodes as 20%, and set the value of *M* as 4. We can simulate the proposed MP-PSO algorithm and analyze its positioning accuracy and positioning coverage rate. In this paper, we select the average error as an indicator to evaluate the performance of localization method [23], and it can be calculated as Equation (15):

$$\overline{Error} = \sum_{i=1}^{K} \frac{|U_{i'} - U_i|}{K \times R_a} \times 100\% \quad (15)$$

where, K is the number of unknown nodes, U_i is the real position of the unknown node, U_i' is the located position of the unknown node, and R_a is the communication radius of the unknown node. The error calculating as this manner denotes the proportion of the distance between the real position and the positioning position to the communication radius of the unknown node.

The localization coverage rate, which denotes the proportion of the effectively located nodes to all the unknown nodes in the network is another important indicator to evaluate the localization algorithm. In this paper, we define the nodes whose error is less than 50% as the effectively located nodes. With the settings above, the localization results of the underwater nodes moving with the water current after 100 s are shown in Figure 8.

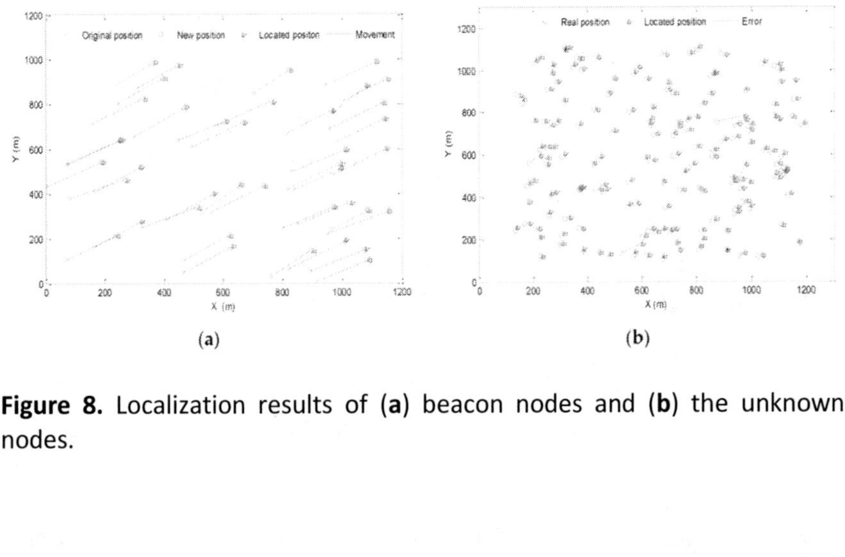

Figure 8. Localization results of (**a**) beacon nodes and (**b**) the unknown nodes.

As shown in Figure 8a, the beacon nodes move with the water current for 100 s, we use the range-based PSO algorithm to locate them and the average error can be calculated, which is 3.26%, that is to say the absolute error is 6.52 m. This means the positioning accuracy of beacon nodes is not bad for this large-scale deployed underwater network. The higher positioning accuracy of the beacon nodes helps to make the error of prediction velocity of unknown nodes smaller, and it ensures the positioning accuracy of the whole algorithm to be relatively ideal. Figure 8b shows the comparison of the located and the real position after the unknown nodes move for 100 s, and it can be seen that some nodes' error is too large, making them not effectively located nodes. The average error of the unknown nodes is 19.4% and the localization coverage rate is 97%.

In the process of the nodes' movement with the water currentw, we can record the position of a certain node at every moment within 100 s to plot its real movement track in 100 s, and we can also plot the located track according to the located position. Selecting the beacon node whose original coordinates are (452.9, 638.8) and the unknown node whose original coordinates are (219.7, 950.2) as the experimental subject, we can plot their real tracks and located tracks, which are shown in Figure 9.

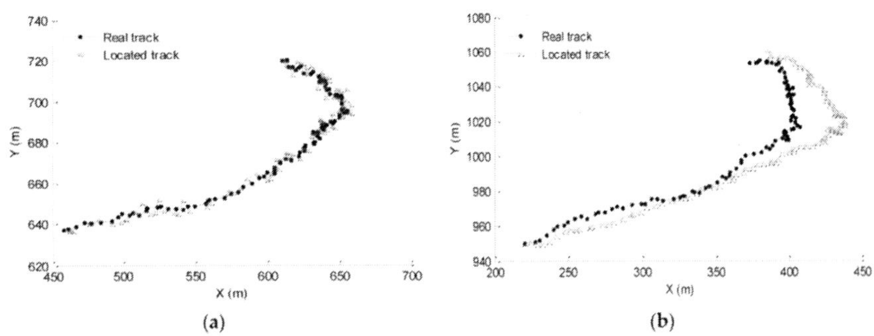

Figure 9. The comparison of (**a**) the beacon node's track and (**b**) the unknown node's track.

Figure 9a shows the comparison between the real and located tracks for the beacon node within 100 s. It can be seen that the localization error almost does not change with the time. That is because the buoys' coordinates are absolutely accurate, and there is no error accumulation in the range-based PSO algorithm. It also indicates that the PSO algorithm has better stability. Figure 9b shows that the error of unknown nodes increases gradually with time. This is because the localization of unknown nodes is a continuous prediction process, which may cause error accumulation and make the located tracks deviate from the real tracks as time goes on. While it is can be seen that the accuracy of the unknown nodes is relatively in the ideal range, the error is always limited within 50% in the process.

4.3. Impacts of Beacon Nodes' Proportion on Localization Results

Because the localization accuracy of beacon nodes is obviously higher than that of the unknown nodes, the proportion of beacon nodes can directly affect the localization results. Keeping the settings in Table 1 and changing the proportion of beacon nodes to 10%, 15%, 20%, 25% and 30%, respectively, the optimal value of the number of referenced nodes is selected according to Figure 7, and we can simulate and compare the localization results of the MP-PSO method, SLMP method and MCL-MP method after 100 s, respectively. The compared results are shown in Figure 10.

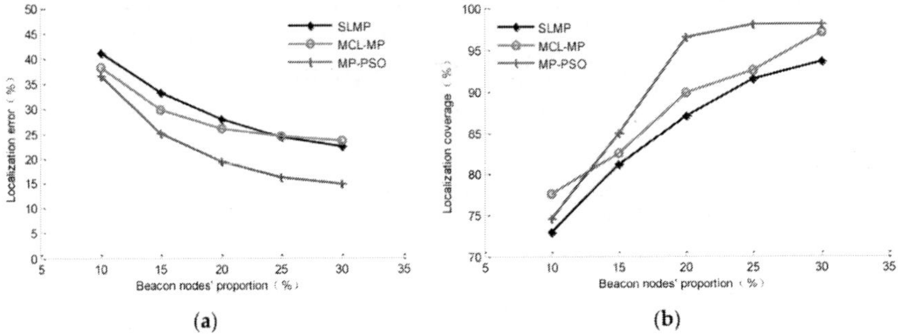

Figure 10. Impacts of the beacon node proportion on localization results: (**a**) on localization error; (**b**) on localization coverage.

Figure 10a shows that MP-PSO method has an obvious localization accuracy advantage compared to the MCL-MP and SLMP methods. When the beacon node proportion is 20%, the average localization error of MP-PSO is decreased 8.7% compared to the SLMP method and 7.1% compared to the MCL-MP method. In addition, we can see that at the beginning of the increase of the proportion of beacon nodes, the localization accuracies of the three methods are all improved obviously. When the beacon node proportion is more than 20%, the decreasing trend of the localization error gradually becomes flat. Therefore, the effect of an excessive increase in the number of beacon nodes is insignificant, and it will only serve to increase the cost of the network, so it is necessary to keep a proper balance between the localization accuracy and the network cost in the deployment of a sensor network.

Figure 10b shows the state of localization coverage rate change with the proportion of beacon nodes for the three methods. As shown in the figure, when the proportion of beacon nodes is more than 20%, the localization coverage rate of the MP-PSO method can be higher than 95%, which is much better than the other two methods. That is because the MP-PSO method chooses the reference nodes with higher confidence coefficient to estimate the unknown nodes' velocities, which can effectively decrease the localization error. In the MCL-MP method, the unknown nodes use the beacon nodes within their one-hop and two-hops range to predict their next position, so even if the proportion of beacon nodes increases more, inevitably some of the nodes cannot be located effectively due to the large localization error. In the SLMP method, the velocities of beacon nodes are estimated by the Durbin algorithm, and the error caused in this process is not related to the beacon node proportion, so its localization coverage rate does not increase as much as the MP-PSO method.

4.4. Impacts of Localization Period T on Localization Results

The localization period *T* is another factor which can affect the localization results. Keeping the settings in Table 1, choosing the proportion of beacon nodes as 20%, and changing the localization period from 1 s to 6 s, respectively, the localization results are as shown in Figure 11.

As shown in Figure 11a, as the localization period increases, the localization accuracy of the three methods decreases faster and faster. The accuracy of the MCL-MP method is affected by the localization period most. This is because this method realizes node localization by sampling prediction and filtering. The sampling range will increase as the localization period increases, which may cause larger localization errors. For the SLMP and MP-PSO methods, the bigger localization period will cause a larger displacement of the node between two positioning moments, and the node's velocity changes all the time, so the error will increase when we use the velocity of one moment to represent the velocity of a period of time. The longer the time is, the larger the error will be. From Figure 11a, we can see that when the localization period is less than 4 s, the error of SLMP and MP-PSO methods will not increase much, but when it continues to increase more than 4 s, the accuracies decrease obviously, so the localization period of these two methods cannot be set too long. Considering the higher calculation efficiency of the MP-PSO method compared to the other methods, the small localization period is acceptable. In addition, Figure 11b shows that when the localization period is less than 4 s, the MP-PSO method has a far higher localization coverage rate than the other methods, and the localization coverage rate will not change much as the localization period changes, which proves that the proposed method has a better stability in this stage.

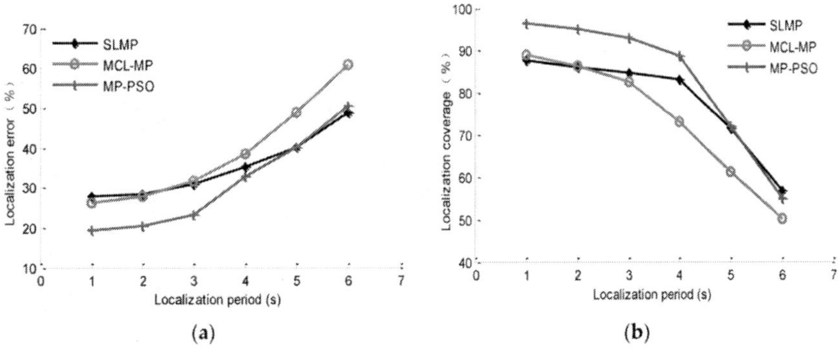

Figure 11. Impacts of localization period on localization results: (**a**) impact on localization error; (**b**) impact on localization coverage.

4.5. Impacts of Nodes' Velocity v on Localization Results

Keeping the parameter settings in Table 1, setting the proportion of beacon nodes as 20%, and changing the nodes' velocity, randomly taking the unknown node whose original coordinate is (229.78, 953.04) as example, the comparisons of tracks under different velocities are shown in Figure 12.

As shown in Figure 12, with different velocities, the deviation of the located and real position of the unknown node becomes large as the time goes on, and this is accord with the analysis of the localization error of unknown nodes above. At the same time, as the velocity increases, we can see that the localization error increases gradually. From Figure 12a to Figure 12d, with the velocity increasing from 1 m/s to 4 m/s, the range of the nodes movement increases as well, and the deviation of the located trajectory from the actual trajectory also increases gradually. That is because when the velocity increases, the distance of nodes' movement between two positioning moments will increase with the same localization period, so the relevant localization error will increase as well. Therefore, this method is suitable for the underwater environment in which the water flow is steady and the mobility of nodes is not very fast. Considering that underwater WSNs are mainly used in the seashore environment currently, where the water is shallow and the nodes' velocity is relatively slow, this method has better applicability in underwater WSNs with mobile nodes.

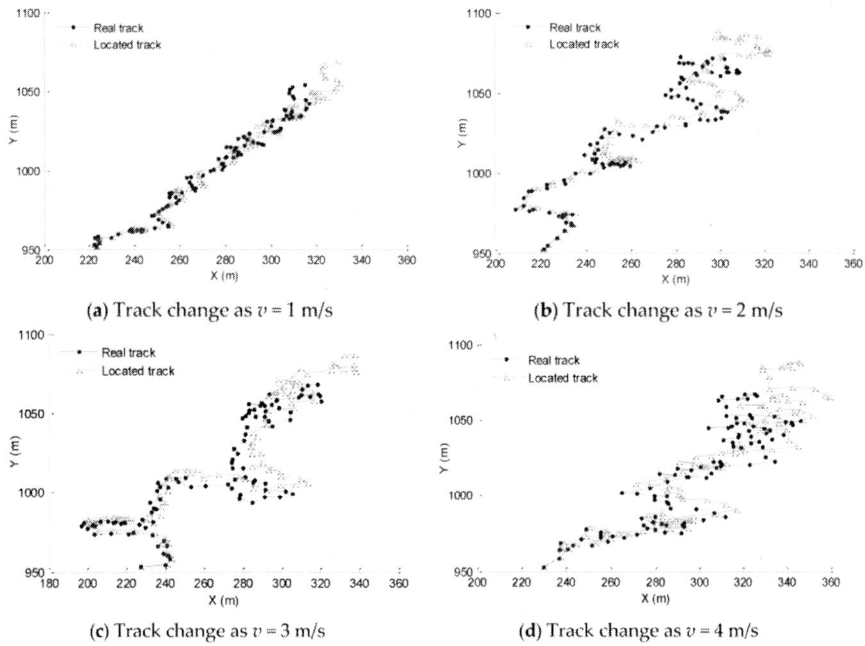

(a) Track change as $v = 1$ m/s
(b) Track change as $v = 2$ m/s
(c) Track change as $v = 3$ m/s
(d) Track change as $v = 4$ m/s

Figure 12. Impacts of nodes' velocity on localization results.

4.6. Analysis of the Energy Consumption

Energy consumption is another important indicator to evaluate the localization methods, and it is mainly the result of the communication activities among the nodes. The communication consumption can be described by the average communication cost, which is defined as the ratio of the overall messages exchanged in the network to the number of effectively localized sensor nodes. Keeping the parameter settings in Table 1, setting the beacon nodes' proportion as 20% and the value of *M* as 4, we can simulate and analyze the average communication cost of the range-based PSO and MP-PSO methods when the number of the whole network nodes changes to 100, 150, 200, 250, 300, 350 and 400, respectively. The result is shown in Figure 13.

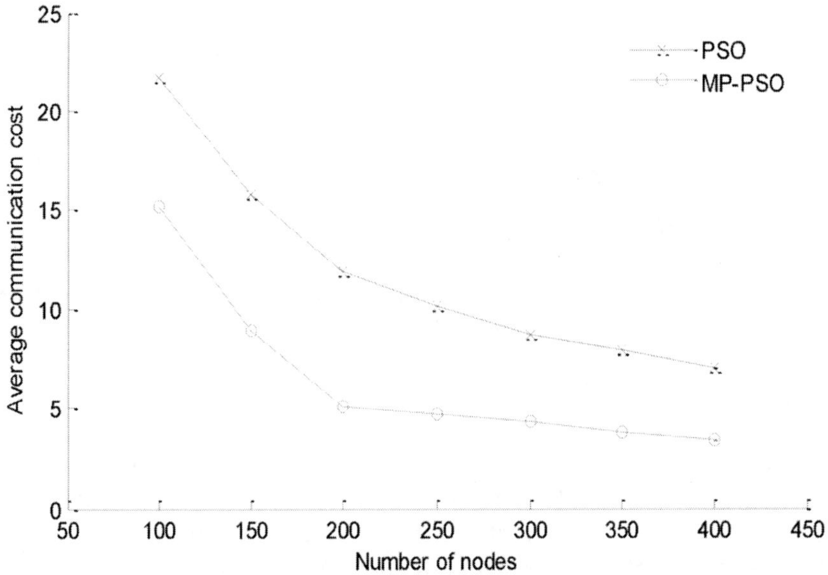

Figure 13. The comparison of energy consumption.

Figure 13 clearly shows that with an increasing number of nodes in the network, the average communication cost of PSO and MP-PSO all decrease at a ratio of about 10% before the number of nodes reaches 200. This is because as the number of nodes increases, the number of effectively located nodes increases drastically. Further, when the number of nodes increases to more than 200, the rates of change of the average communication cost for the two methods all become slow, and ultimately stabilize at some small value. At this stage, the droop rates of PSO and MP-PSO are about 2.25% and 0.6% respectively. Meanwhile, Figure 13 also shows the mobility prediction scheme can greatly reduce the average communication cost of the network. When the number of nodes is less than 250, the average communication cost of MP-PSO is approximate 6 lower than PSO, and when the number of nodes is more than 250, the average communication cost of MP-PSO is approximately 4 lower than PSO. The decrease of average communication cost of the MP-PSO method is about 50% compared to the PSO method. This is because in the mobility prediction procedure, the unknown nodes need not establish communication links with every beacon node and the buoys. In fact, the unknown nodes only need to communicate with several reference nodes to predict their next location. The proposed scheme can effectively reduce the overall data package exchange in the network. Accordingly, the average communication cost of the network will be reduced. This is quite significant for UWSNs with limited bandwidth and energies.

5. CONCLUSIONS

Focusing on the problems of underwater WSNs, like sparse deployment and mobility of the nodes driven by water currents, a multi-step localization method based on mobility prediction and PSO algorithm is proposed. The beacon nodes are located by the range-based PSO algorithm and the velocities can be calculated by using the localization results, and then the unknown nodes are located by the mobility prediction based on the estimation of the spatial correlation of underwater objects' mobility. The simulation results show that this method can provide highly accurate localization of beacon nodes, and the positioning accuracy and positioning coverage rate can be kept at a better level. Compared to the SLMP method which uses the Durbin algorithm to estimate the beacon nodes' velocity and the MCL-MP method which is based on sampling prediction and filtering, with the same localization period, the MP-PSO method obviously has higher localization accuracy and coverage rate. With the same proportion of beacon nodes, the localization result of MP-PSO also has obvious advantages with a shorter localization period as well. In addition, the energy consumption can be reduced by the proposed mobility prediction scheme, so the proposed localization method of mobile nodes in UWSNs can be applicable in the fields of military defense, environmental monitoring and animal tracking, *etc.*

ACKNOWLEDGMENTS

This work was supported by National Nature Science Foundation of China (No. 61273068), and International Exchanges and Cooperation Projects of Shanghai Science and Technology Committee (No. 15220721800).

AUTHOR CONTRIBUTIONS

Ying Zhang conceived and designed the research and experiments, and contributed as the lead author of the article; Ying Zhang and Jixing Liang wrote the article; Jixing Liang performed the experiments; Ying Zhang and Jixing Liang contributed to revising and proofreading of the article; Shengming Jiang gave suggestion and analyzed the data; Wei Chen gave more valuable suggestion to the research, and audited the article.

CONFLICTS OF INTEREST

The authors declare no conflict of interest.

REFERENCES

1. Heidemann, J.; Stojanovic, M.; Zorzi, M. Underwater sensor networks: applications, advances and challenges. *Philos. Trans. R. Soc. A Math. Phys. Eng. Sci.* **2012**, *370*, 158–175.

2. Javaid, N.; Jafri, M.R.; Khan, Z.A.; Alrajeh, N.; Imran, M.; Vasilakos, A. Chain-based communication in cylindrical underwater wireless sensor networks. *Sensors* **2015**, *15*, 3625–3649.

3. Climent, S.; Sanchez, A.; Capella, J.V.; Meratnia, N.; Serrano, J.J. Underwater acoustic wireless sensor networks: advances and future trends in physical, MAC and routing layers. *Sensors* **2014**, *14*, 795–833.

4. Liu, Z.; Gao, H.; Wang, W.; Chang, S.; Chen, J. Color filtering localization for three-dimensional underwater acoustic sensor networks. *Sensors* **2015**, *15*, 6009–6032.

5. Beniwal, M.; Singh, R. Localization techniques and their challenges in underwater wireless sensor networks. *Int. J. Comput. Sci. Inf. Technol.* **2014**, *5*, 4706–4710.

6. Krishna, C.R.; Yadav, P.S. A hybrid localization scheme for underwater wireless sensor networks. In Proceedings of the IEEE 2014 International Conference on Issues and Challenges in Intelligent Computing Techniques (ICICT), Ghaziabad, India, 7–8 February 2014; pp. 579–582.

7. Erol-Kantarci, M.; Mouftah, H.T.; Oktug, S. A survey of architectures and localization techniques for underwater acoustic sensor networks. *IEEE Commun. Surv. Tutor.* **2011**, *13*, 487–502.

8. AlHajri, M.I.; Goian, A.; Darweesh, M.; AlMemari, R.; Shubair, R.M.; Weruaga, L.; Kulaib, A.R. Hybrid RSS-DOA technique for enhanced WSN localization in a correlated environment. In Proceedings of the 2015 IEEE International Conference on Information and Communication Technology Research (ICTRC), Abu Dhabi, United Arab Emirates, 17–19 May 2015; pp. 238–241.

9. El Assaf, A.; Zaidi, S.A.R.; Affes, S.; Kandil, N. Range-free localization algorithm for heterogeneous wireless sensor networks. In Proceedings of the 2014 IEEE Wireless Communications and Networking Conference (WCNC), Istanbul, Turkey, 69 April 2014; pp. 2805–2810.

10. Hu, L.; Evans, D. Localization for mobile sensor network. In Proceedings of the ACM 10th annual international conference on Mobile computing and networking, Philadelphia, PA, USA, 26 September–1 October 2004; pp. 45–57.

11. Soltaninasab, B.; Sabaei, M.; Amiri, J. Improving Monte Carlo localization algorithm using time series forecasting method and dynamic sampling in mobile WSNs. In Proceedings of the 2010 Second International Conference on Communication Systems, Networks and Applications, Hong Kong, China, 29 June–1 July 2010; pp. 389–396.

12. Alcocer, A.; Oliveira, P.; Pascoal, A. Study and implementation of an EKF GIB-based underwater positioning system.*Control Eng. Pract.* **2007**, *15*, 689–701.

13. Cheng, X.; Shu, H.; Liang, Q.; Du, D.H.C. Silent positioning in underwater acoustic sensor networks. *IEEE Trans. Veh. Technol.* **2008**, *57*, 1756–1766.

14. Zhou, Z.; Peng, Z.; Cui, J.H.; Shi, Z.; Bagtzoglou, A.C. Scalable localization with mobility prediction for underwater sensor networks. *IEEE Trans. Mob. Comput.* **2011**, *10*, 335–348.

15. Cheng, W.; Teymorian, A.Y.; Ma, L.; Cheng, X.; Lu, X.; Lu, Z. Underwater localization in sparse 3D acoustic sensor networks. In Proceedings of the 27th IEEE Conference on Computer Communications INFOCOM, Phoenix, AZ, USA, 13–18 April 2008; pp. 236–240.

16. Yang, H.; Sikdar, B. A Mobility Based Architecture for Underwater Acoustic Sensor Networks. In Proceedings of the IEEE Global Telecommunications Conference (GLOBECOM), New Orleans, LO, USA, 30 November–4 December 2008; pp. 5107–5111.

17. Pu, L.; Luo, Y.; Mo, H.; Le, S.; Peng, Z.; Cui, J.H.; Jiang, Z. Comparing underwater MAC protocols in real sea experiments. *Comput. Commun.* **2015**, *56*, 47–59.

18. Kuperman, W.A.; Roux, P. Underwater Acoustics. In *Springer Handbook of Acoustics*; Springer: New York, NY, USA, 2014; pp. 157–212.

19. Guo, Z.; Luo, H.; Hong, F.; Yang, M.; Lionel, M. Current progress and research issues in underwater sensor networks.*J. Comput. Res. Dev.* **2010**, *47*, 377–389.

20. Singh, S.; Shivangna, S.; Mittal, E. Range based wireless sensor node localization using PSO and BBO and its variants. In Proceedings of the 2013 IEEE International Conference on Communication Systems and Network Technologies (CSNT), Gwalior, India, 6–8 April 2013; pp. 309–315.

21. Zain, I.F.M.; Shin, S.Y. Distributed localization for wireless sensor networks using binary particle swarm optimization (BPSO). In Proceedings of the 2014 IEEE 79th Vehicular Technology Conference (VTC Spring), Seoul, Korea, 18–21 May 2014; pp. 1–5.

22. Pluhacek, M.; Senkerik, R.; Zelinka, I. Particle swarm optimization algorithm driven by multichaotic number generator. *Soft Comput.* **2014**, *18*, 1–9.

23. Kulkarni, R.V.; Forster, A.; Venayagamoorthy, G.K. Computational intelligence in wireless sensor networks: a survey.*IEEE Commun. Surv. Tutor.* **2011**, *13*, 68–96.

Chapter 8

Fuzzy Honey Bees Foraging Optimization: Swarm Intelligence Approach for Clustering

Ali-Asghar Gholami1, Ramin Ayanzadeh2 and Elaheh Raisi3

[1]Department of Applied Mathematics, Parand Branch, Islamic Azad University, Parand, Iran

[2]Department of Computer, Bojnord Branch, Islamic Azad University, Parand, Iran

[3] Department of Computer, Science and Research Branch, Islamic Azad University, Parand, Iran

ABSTRACT

Clustering is one of the most important steps in data mining; it is known for its phenomenal functionalities in complex real world applications including biology, basic science, medicine, engineering and social science. In this sense, owing to the remarkable effects of clustering on data mining area, wide varieties of clustering approaches have been introduced to cluster data into significant subsets in order to obtain useful information. In this study, a novel clustering method based on honey bees foraging optimization algorithm and fuzzy rules is proposed. In the proposed method, fine shade of local and global search in honey bees optimization algorithm is schemed to be applied to improve the clustering efficiency.

Furthermore, fuzzy operators are employed to enhance the performance of new proposed approach and prevent premature convergence. To verify and validate the functionality proposed of method, new method is run on three known data sets of the UCI Machine Learning Repository. Results of clustering reveal that proposed method estimate more desirable clusters compared to the state of the art clustering methods. Moreover, this method appears very stable in multiple tests.

KEYWORDS

Clustering, data-mining, fuzzy sets, honey bees foraging optimisations, meta-heuristics, pattern recognition

1. INTRODUCTION

Data mining, a synonym to "data knowledge discovery", is an analytical process of exploring data (usually large amounts of data) from various perspectives and summarizing it so that we can focus on the most important information. Data mining is also useful to discover the various patterns between the data and establish the relations between them. Data mining techniques are used in wide variety of research areas including but not limited to biology, medicine, mathematics, cybernetics, engineering, economics and marketing (Ayanzadeh et al., 2011a; Webb, 2002). Data mining problems including regression, prediction, classification, clustering, pattern mining, feature selection and feature extraction have been subject of extensive research for several decades (Theodorodis and Koutroumbus, 2006; Webb, 2002). Clustering and classification are two significant and frequently used techniques which are extremely powerful in their respective goals (Ayanzadeh et al., 2011a; Theodorodis and Koutroumbus, 2006; Webb, 2002).

In the terminology of data mining, the core objective of classification and clustering is to utilize a data set to train a system. Although classification and clustering are often mentioned in the same breath, they are different analytical methods. Most of the learning algorithms used in data mining are either supervised learning or unsupervised learning. Clustering is one form of unsupervised learning and involves grouping data into categories based on some measure of inherent similarity such a way that objects in the same category are more similar to each other than to those in other categories. Classification is the corresponding supervised procedure. In classification, a set of predefined categories exists and the problem is identifying to which categories the new observation belongs (Theodorodis and Koutroumbus, 2006; Webb, 2002).

Clustering algorithms are sorted out into two major categories with respect to determining the number of clusters in data set. In first group, the number of clusters is assumed to be a fixed parameter which is specified in advance by the analyst and the clusters are formed from either an initial guess at the cluster centres, or from random initial conditions. On the contrary, in the second group, the number of clusters is not specified beforehand. These clustering algorithms detect the number of clusters concerning the structure and nature of training data set (Theodorodis and Koutroumbus, 2006; Webb, 2002).

There are large number of clustering methods introduced in the literature. Hierarchical clustering algorithms are renowned clustering methods used in wide range of applications in which nested clusters are produced. Hierarchical clustering can be either agglomerative, with fewer clusters at the higher level, or divisive which separate the n objects into more and finer groups in sequential steps (Webb, 2002). Due to high computational complexity of hierarchical clustering methods which makes them too slow for large data sets, optimal efficient must be found in order to cluster the large data sets with high dimensionality (Du, 2010; Chen, 2013; Nanda and Panda, 2013; Hongwei et al., 2011). Some of these good methods are single connection hierarchical algorithm, complete connection hierarchical algorithm and sum of squares hierarchical algorithm (Nanda and Panda, 2013; Hongwei et al., 2011).

Although the proposed methods improve the efficiency of clustering, they still suffer from undesirable functionalities in real world scenarios in which very large data sets are supposed to be processed (Nanda and Panda, 2013; Hongwei et al., 2011). To address this issue, intelligent numerical methods are used to handle high complexity and dimensionality of real world applications (Ayanzadeh et al., 2009b; Setayeshi and Fadaei, 2011; Shahamatnia et al., 2011). Evolutionary algorithms have been proven to be a very powerful mechanism in finding good solutions to difficult problems (Hongwei et al., 2011). The flexibility associated with Genetic Algorithms (GA) is one important aspect to take into account. GAs are the most well known intelligent techniques that are applied for clustering problems (Hongwei et al., 2011). With regard to the quality of the solutions that evolutionary algorithms have shown in different types of fields and problems, it makes sense to try to use them in clustering applications more seriously.

Artificial Neural Networks (ANN) are graph based numerical and intelligent approach that have served significant functionalities in data mining more preciously in classification and clustering applications. Owing to the desirable performance of ANN in dealing with large scale data sets, they are widely used for real world clustering applications (Du, 2010).

In the same way, Neuro-Fuzzy systems are also utilized for data mining purposes. In fact, Neuro-Fuzzy systems are ANN based implementation of fuzzy systems that can represent the knowledge more properly. On this basis, ANFIS serves ourstaning performance in real world clustering applications (Chen, 2013). However, it's worth noting that ANFIS can not handle large scale data setes and size reduction pre-processes are required (Engelbrecht, 2007; Jang et al., 2005; Chiou and Lan, 2001). Cellular Automata (CA) and Learning Automata are also being tried for data-mining applications (Torkestani and Meybodi, 2010; Ayanzadeh et al., 2009a; Ayanzadeh et al., 2010).

Honey Bees Foraging Optimization (HBFO) is a novel swarm intelligence approach that can be employed for solving many types of problems. In other words, due to the harmony between local and global search in HBFO, it potentialy can serve phenomenal functionality in real world problem solving (Ayanzadeh et al., 2011b). Furthermore, fuzzy sets provide linguistic variables to represent the knowledge more properly (Ayanzadeh et al., 2012). In this sense, in this study honey bees foraging optimization algorithms are enhanced using fuzzy concepts in evaluating the nectar amount of food sources to propose a novel hybrid intelligent clustering method. From another perspective fuzzy operators are injected into the structure of HBFO algorithm to explore more accurate clusters. Regarding to the stable infrastructure of HBFO and due to the capability of fuzzy linguistic variables in knowledge representation, proposed clustering approach provides more desirable results comparing to other meta-heuristics.

2. HONEY BEES FORAGING OPTIMIZATION

Honey bees are one of the most popular and beneficial insects all over the world thanks to their capabilities to produce honey which is used in various foods and beverages as a sweetener and flavorings and in various medicinal traditions for treatment. Honey bees build their hives in forests, mountains and crack of rocks. Each hive includes vertical parts which are made up of wax. These vertical parts consist of several regular hexagons chambers (Afshar et al., 2007; Pham et al., 2006a). Honey bees are social insects and have complex social life. Each hive might host hundreds or thousands of individuals. There is only one queen in each hive. More interestingly, in honey bee societies, only queens can be fertilized and lay eggs for reproduction purposes. This fertilization is performed during a mating dance which is occurred once in a lifetime (Ayanzadeh et al., 2011b; Sunil and Craig, 2004).

Other than queen and few hundred males, main fraction of the population is unproductive worker bees. For security reasons, queen, males and workers must identify each other not to let stranger insects to come in the hive. This also helps honey bees to communicate, share knowledge about the location of food sources and alert any attacks by foreign bees. All these message passing operations are done by some special dances (Afshar et al., 2007; Ayanzadeh et al., 2011b).

Queen and male bees are responsible only for reproduction. On the other side, worker bees that have no mailting capability are supposed to do a considerable amount of duties including foraging, cleaning, building and maintaining hive, taking care of infant bees and so on. Foraging appears as one of the most important aspects of the system of animate nature which must be undertaken perfectly by workers (Afshar et al., 2007). To this end, honey bees have developed two different responsibilities among worker bees in regards to food production: Scout bees and forager bees. These bees use various "body language" techniques to communicate with one another. Scout bees are selected to search whole environment to seek food sources. Afterward, scout bees return to the hive and share the exploration findings among other worker bees. It is worth to note that scout bees have no initial information about the environment; so, they look for new food sources by chance and stochastically. During the exploration flight, scout bees test the quality of food sources (flowers). Then, food sources are evaluated based on different parameters like quality and amount of nectars, distance from the hive and type of the flowers (Pham et al., 2006b; Haddad and Afshar, 2004).

After the first exploration flight, each scout bee is followed by some worker bees to gather the nectar. Apparently, number of followers is directly related to the quality of food sources that have been found by scouts. That is, those scout bees that find better food sources are followed by more worker bees. Scout bees with the help of worker bees gather nectar of flowers in the area (Ayanzadeh et al., 2011b; Haddad and Afshar, 2004).

Worker bees that accompany the scout bees when they are collecting nectar, exploit the neighbourhood to find better food soutces in that area. Accordingly, if they discover places with better food sources during the exploitation, they would play the role of scout bees. On the other side, worker bees also share their knowledge about the environment when they return to the hive. Thus, these worker bees will be followed by other worker bees in the next flight (Afshar et al., 2007).

Similar to all other bio-inspired optimization algorithms, honey bees foraging optimization is an optimisation algorithm inspired by the natural foraging behaviour of honey bees to find the optimal solution. Both honey bees optimization algorithms (maiting and foraging) are considered swarm intelligence that have been extensively used as search and optimization tools in various problem domains (Ayanzadeh et al., 2011b). Honey bees foraging optimization technique is a computational model of honey bees behaviour when they are searching for food sources (Afshar et al., 2007).

In this method, solutions are assumed as food sources and artificial honey bees explore and exploit the search spave to find optimum solution. In the same way, with traditional meta-heuristic methods, such as genetic algorithms, potential solutions are formulated by vectors of numbers. These food sources are initialized randomly usually with uniform distribution (Afshar et al., 2007; Ayanzadeh et al., 2011b).

Afterward, iterative optimization process is begun. The iteration process will continue until one of the criterias for termination is met. Food sources are evaluated based on nectar amount (fitness value). Afterward, food sources are categorized into two desired and undesired groups. From another perspective, first K best food sources are assumed as desired and rests are called undesired (Ayanzadeh et al., 2011b; Jaberi et al., 2012).

Desirable food sources are obtained by a local search algorithm like hill climbing or simulated annealing heuristics and in the opposite side, undesirable food sources are explored by global search techniques. In both local and global searches, when a better food source is found, it is substituted with primary food sources. In better words, initial food sources can be replaced by the output of exploration and exploitation search (Ayanzadeh et al., 2011b; Jaberi et al., 2011; Khosravani-Rad et al., 2014; Marzban et al., 2014).

Local and global searches will be conducted until one termination condition is met. Termination condition might be either the number of iterations or finding a food source with predefined nectar amount. The number of worker bees that follow scout bees during exploring has a direct relationship with the quality of food sources that have been found by scouts. Parameters of local and global search algorithms in standard simulations of honey bees foraging optimization are supposed to be the same for desirable and undesirable food sources (Ayanzadeh et al., 2011b; Haddad and Afshar, 2004).

3. FUZZY HONEY BEES FORAGING OPTIMIZATION BASED CLUSTERING

As mentioned earlier, owing to the important functionality of clustering in solving real world data processing problems, wide variety of statistical and intelligent clustering methods have been introduced. On this basis, different branches of computer science and artificial intelligence including hierarchical clustering algorithms, Genetic Algorithm, Particle Swarm Optimization, Artificial Neural Networks and Neuro-Fuzzy systems are applied for clustering purposes. In this section, a novel hybrid clustering method based on honey bees foraging optimization and fuzzy sets is proposed.

Structure: Proposed clustering method is founded on honey bees foraging optimization algorithm. Indeed, this method is employed for clustering with pre-defined and fixed number of clusters. This algorithm seeks problem state space to find optimum clusters' canters. Cluster numbers, as a system parameter specifies the size of food sources. In order to find the optimum positions of C clusters in a problem in Rn state space, food sources will consist of Cxn elements. Food sources are initialized by random values. Then, food sources are evaluated in each iteration using nectar amount function which is performed similar to fitness function in genetic algorithms (Ayanzadeh et al., 2011b; Jaberi et al., 2013).

More specifically, food sources are divided into C groups in which each group represents the position of a cluster. Apparently, in this sense, the clustering problem is considered as an optimization problem. Thus, honey bees foraging optimization algorithm is supposed to find the minimum error (maximum likelihood) of nectar amount function). Undoubtedly, nectar amount function serves key functionality and changes an optimization process to a clustering paradigm.

Nectar amount function: The main difference between proposed method and standard honey bees foraging optimization lies in examining the amount of nectar. To this end, fuzzy operators are employed to map qualitative values into a quantitative scale. In other words, linguistic variables are used to enhance the evaluation of nectar amount.

In traditional clustering methods, each input pattern is assigned to be blong to the nearest cluster. In proposed method, however, input patterns are belong to all clusters with different membership rates. To achieve this goal, Gaussian fuzzifier is utilized to calculate the membership values. For example, to assess amount of nectar in vector X1x6 in R3 state space, three cluster centers are represented as food sources. Thus, membership value of belonging X to C1, C2 and C3 are calculated by Eq. 1:

$$\mu_j = \frac{e^{\left(\frac{(X_i - C_t)^2}{-2\sigma^2}\right)}}{\sqrt{2\pi\sigma^2}} \tag{1}$$

where, Xi is ith pattern and Cj represents the centre of jth cluster. After calculating Euclidean distance between training data and centres of clusters, ith input sample will be assigned to the cluster with maximum nectar amount (maximum membership value).

Apparently, this type of clustering is completely different from fuzzy clustering aspects. More narrowly, multi-belonging is taken into account in evaluating the amounts of nectar and every pattern is finally assigned to a unique cluster.

Total clustering error for each cluster is summation of Euclidean distances between each pair of patterns in the specified cluster (total internal distances). Finally, clustering error for an instance food source will be the summation of all corresponding clusters' errors (internal distances). Using this technique, the algorithm will be converged to food sources with minimum clustering error.

4. EXPERIMENTAL RESULTS

In this section, results of implementing and testing the proposed clustering algorithm along with its comparison with the results of an enhanced version of Genetics Algorithm are mentioned. In this work, three clustering problems of UCI data repository are used to evaluate the performance of algorithm. The selected datasets are Iris, Glass and Wine in which, all patterns have class title.

Iris: In this test, Iris data set is used because of its familiarity and remarkable fame in classification and clustering methods. Nearly every data mining package comes with this data set. The Iris data set consists of three types of plants which are class labels (Setosa, Versicolor and Virginica), with 50 instances per each class, represented by 4 features.

In this experiment, patterns are categorized into three clusters. Clustering is performed using proposed method and an enhanced version of genetic algorithm. Figure 1 and 2 illustrate convergence of genetic algorithm and fuzzy honey bees foraging optimization clustering method for clustering, respectively.

Maximum clustering errors are also given in Table 1. It is obvious that both enhanced genetic algorithm and fuzzy honey bees foraging optimization clustering techniques demonstrate similar behaviour in clustering Iris data set.

Glass: In this simulation another benchmark dataset from UCI repository named Glass is chosen to evaluate the functionality of proposed clustering method. The Glass dataset has 10 attributes which are features and 7 types of glasses that here are considered to be class labels.

For this dataset patterns were categorized into 6 clusters. Figure 3 and 4 show convergence of enhanced genetic algorithm and fuzzy honey bees foraging optimization clustering methods for clustering, respectively. Again, maximum clustering errors for both methods are demonstrated in Table 2. As can be seen, the proposed method tends to be more robust and efficient in clustering than enhanced genetic algorithm.

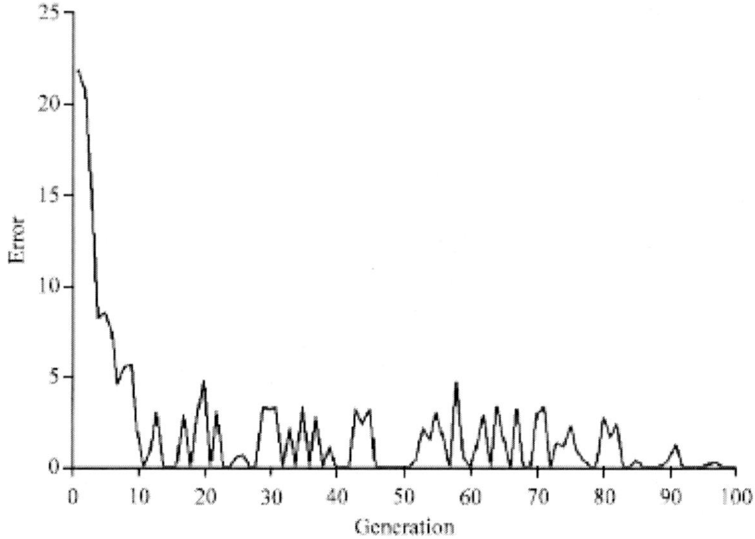

Figure 1: Convergence of enhanced genetic algorithm in clustering of Iris data set

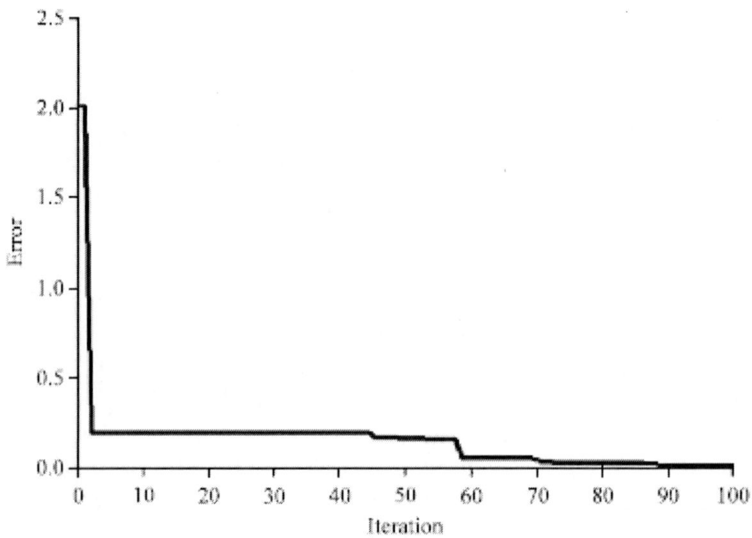

Figure 2: Convergence of fuzzy honey bees foraging optimization clustering in clustering of Iris data set

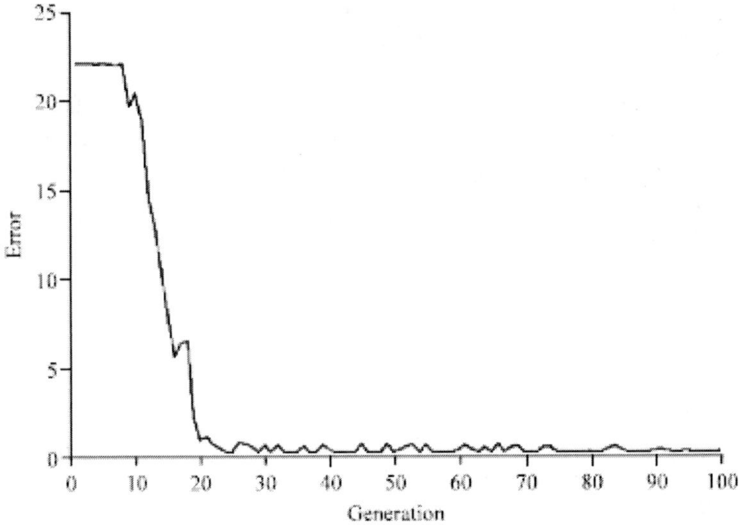

Figure 3: Convergence of enhanced genetic algorithm in clustering of Glass data set

Table 1: Maximum clustering errors obtained by enhanced GA and FHBFO methods for Iris data set

set

GA	FHBFO
0.0018	0.0018

Table 2: Maximum clustering errors obtained by enhanced GA and FHBFO methods for Glass data set

set

GA	FHBFO
0.0851	0.0512

Wine: Performance of proposed algorithm was tested based on Wine data set from UCI repository. The decision for choosing this database was motivated by the fact that this dataset has greater number of dimensions compare to Iris and Wine. In this way, validation is carried out on more complicated data set. Wine dataset is known to have 3 classes with 13 features and 187 numbers of instances.

In this evaluation, patterns are supposed to be clustered into three groups. Figure 5 and 6 depict convergence of enhanced genetic algorithm and the proposed hybrid method for clustering of Wine data set.

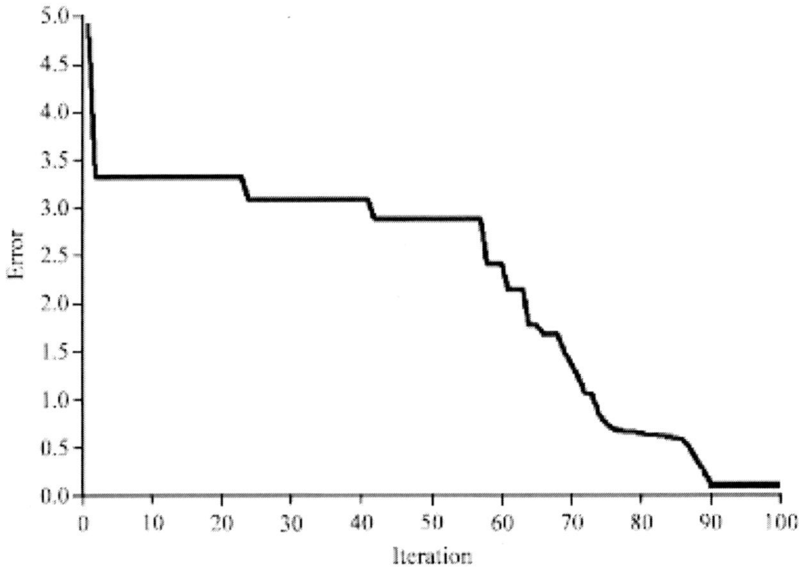

Figure 4: Convergence of fuzzy honey bees foraging optimization clustering in clustering of Glass data set

5. CONCLUSION

Considering the importance of clustering techniques for exploratory data analysis in solving real world applications, in this study a novel hybrid method for clustering was proposed. In this method fuzzy operators were injected into the infrastructure of honey bees foraging optimization algorithm. In some sense, clustering challenges are formulated as optimization problems.

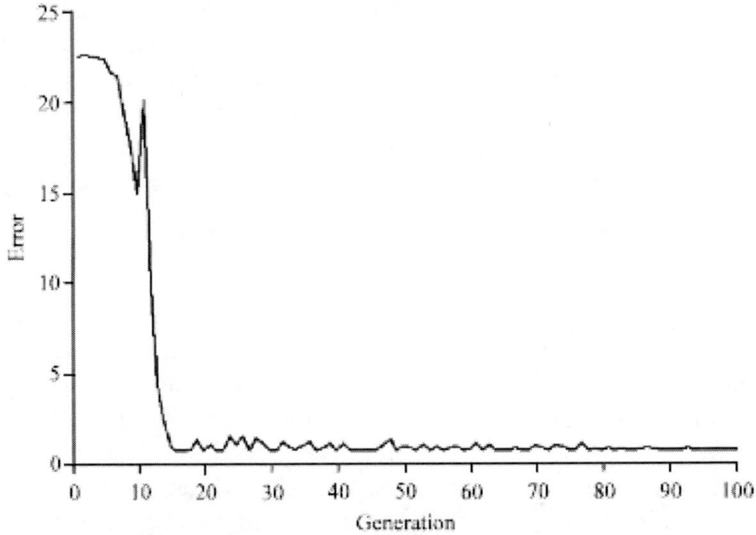

Figure 5: Convergence of enhanced genetic algorithm in clustering of Wine data set

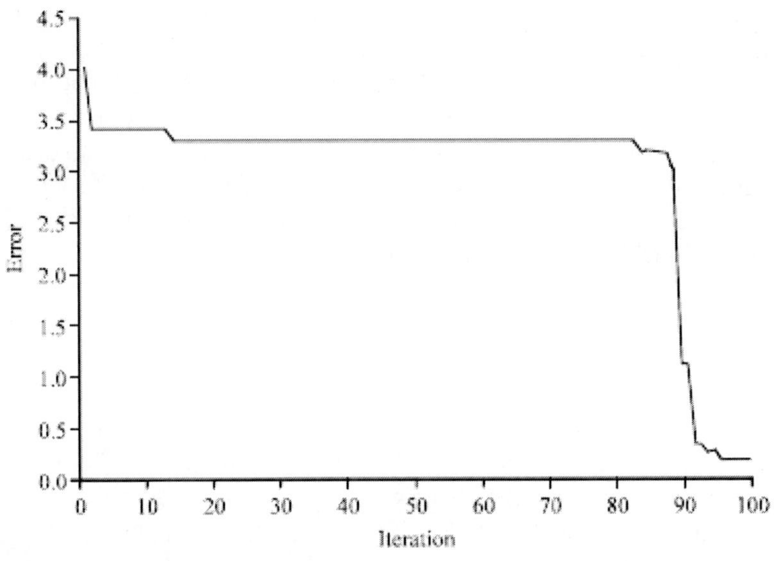

Figure 6: Convergence of fuzzy honey bees foraging optimization clustering in clustering of Wine data set

Table 3 presents the maximum clustering errors of clustering the Wine data set using both aforementioned methods. Apparently, fuzzy honey bees foraging optimization clustering serves more desirable functionality and provides more accurate clusters than enhance genetic algorithm.

set

GA	FHBFO
0.2314	0.0996

In this proposed method, nectar amount function which serves same functionality as fitness function in genetic algorithms, is in charge of evaluating clustering error to convert clustering problem in to optimization problem. In other words, food sources represent centres of clusters and amount of nectar was supposed to be total clustering error. From another attitude, for each food source, nectar amount function calculates the summation of euclidean distances between every two patterns for all patterns in a cluster and clustering error was assumed as total value of internal distances. Furthermore, fuzzy operators were utilized to map qualitative linguistic variables into quantitative values.

To test the efficiency and reliability of our proposed clustering method, several experiments were conducted with three datasets from the UCI database ranging from a small number of classes and small-dimensional features to a large number of classes and large-dimensional features. The experiments were intended to assess the clustering accuracy of the proposed method and compare the results with an enhanced version of genetic algorithms. The experimental results show that the proposed hybrid method produces more accurate and higher quality clusters in comparison with an enhanced genetic algorithm.

REFERENCES

1. Afshar, A., O.B. Haddad, M.A. Marino and B.J. Adams, 2007. Honey-bee mating optimization (HBMO) algorithm for optimal reservoir operation. J. Franklin Inst., 344: 452-462.

2. Ayanzadeh, R., A.S.Z. Mousavi and E. Shahamatnia, 2012. Fuzzy cellular automata based random numbers generation. Trends Applied Sci. Res., 7: 96-102.

3. Ayanzadeh, R., A.S.Z. Mousavi and H. Navidi, 2011. Honey bees foraging optimization for mixed Nash equilibrium estimation. Trends Applied Sci. Res., 6: 1352-1359. |

4. Ayanzadeh, R., A.S.Z. Mousavi and S. Setayeshi, 2011. Fossil fuel consumption prediction using emotional learning in Amygdala. Proceedings of the 19th Iranian Conference on Electrical Engineering, May 17-19, 2011, Tehran, Iran, pp: 1-6.

5. Ayanzadeh, R., E. Shahamatnia and S. Setayeshi, 2009. Determining optimum queue length in computer networks by using memetic algorithms. J. Applied Sci., 9: 2847-2851.

6. Ayanzadeh, R., K. Hassani, Y. Moghaddas, H. Gheiby and S. Setayeshi, 2009. Innovative approach to generate uniform random numbers based on a novel cellular automata. J. Applied Sci., 9: 4071-4075.

7. Ayanzadeh, R., K. Hassani, Y. Moghaddas, H. Gheiby and S. Setayeshi, 2010. Multi-layer CA for normal random number generation. Proceedings of the 18th Iranian Conference on Electrical Engineering, May 11-13, 2010, Isfahan, Iran -.

8. Chen, M.Y., 2013. A hybrid ANFIS model for business failure prediction utilizing particle swarm optimization and subtractive clustering. Inform. Sci., 220: 180-195.

9. Chiou, Y.C. and L.W. Lan, 2001. Genetic clustering algorithms. Eur. J. Operat. Res., 135: 413-427.

10. Du, K.L., 2010. Clustering: A neural network approach. Neural Networks, 23: 89-107.

11. Engelbrecht, A.P., 2007. Computational Intelligence: An Introduction. 2nd Edn., John Wiley and Sons Ltd., New York, USA., ISBN-13: 9780470035610, Pages: 597.

12. Haddad, B. and A. Afshar, 2004. MBO (Marriage Bees Optimization), A new heuristic approach in hydro systems design and operation. Proceedings of 1st International Conference on Managing Rivers in the 21st Century: Issues and Challenges, September 21-23, 2004, Penang, Malaysia, pp: 499-504.

13. Hongwei, Z., Y Zhenyu and Z Shurong, 2011. Multi-objective supervised clustering GA and wintry climate forecast. Proc. Eng., 15: 1580-1584.

14. Jaberi, A., R. Ayanzadeh and A.S.Z. Mousavi, 2012. Two-layer cellular automata based cryptography. Trends Applied Sci. Res., 7: 68-77.

15. Jaberi, A., R. Ayanzadeh and E. Shahamatnia, 2011. A novel fuzzy cellular automata for uniform random number generation. Proceedings of the 2nd International Conference on Contemporary Issues in Computer and Information Sciences, May 31-June 2, 2011, Zanjan, Iran.

16. Jaberi, A., R. Ayanzadeh, E. Raisi and A. Sadighi, 2013. Central limit theorem based cellular automata for generating normal random numbers. Inform. Technol. J., 12: 2440-2446.

17. Jang, J.S.R., C.T. Sun and E. Mizutani, 2005. Neuro-Fuzzy and Soft Computing: A Computational Approach to Learning and Machine Intelligence. Prentice Hall of India, New Dehli, India.

18. Khosravani-Rad, A., R. Ayanzadeh and E. Raisi, 2014. Dynamic parameters optimization for enhancing performance and stability of PSO. Trends Applied Sci. Res. (In Press).

19. Marzban, F., R. Ayanzadeh and P. Marzban, 2014. Discrete time dynamic neural networks for pre-dicting chaotic time series. J. Artif. Intell. (In Press).

20. Nanda, S.J. and G. Panda, 2013. Automatic clustering algorithm based on multi-objective Immunized PSO to classify actions of 3D human models. Eng. Appl. Artifi. Intell., 26: 1429-1441.

21. Pham, D.T., A. Ghanbarzadeh, E. Koc, S. Otri, S. Rahim and M. Zaidi, 2006. The bees algorithm-a novel tool for complex optimization problems. Proceedings of the 2nd International Virtual Conference on Intelligent Production Machines and Systems, July 3-14, 2006, Cardiff University, UK., pp: 454-459.

22. Pham, D.T., E. Koc, A. Ghanbarzadeh and S. Otri, 2006. Optimization of the weights of multi-layered perceptrons using the bees algorithm. Proceedings of 5th International Symposium on Intelligent Manufacturing Systems, December 2006, Tehran, Iran, pp: 38-46.

23. Setayeshi, S. and A.H. Fadaei, 2011. CA for Optimization. Nova Science Publishers Inc., New York, USA., ISBN-13: 9781611227437.

24. Shahamatnia, E., R. Ayanzadeh, R.A. Ribeiro and S. Setayeshi, 2011. Adaptive imitation scheme for memetic algorithms. Adv. Inform. Commun. Technol., 349: 109-116.

25. Sunil, N. and T. Craig, 2004. On Honey Bees and Dynamic Server Allocation in Internet Hosting Centers. Sage Publications, New York, USA., pp: 223-240.

26. Theodorodis, S. and K. Koutroumbus, 2006. Pattern Recognition: Introduction. 3rd Edn., Elsevier Academic Press, New York, USA., Pages: 856.

27. Torkestani, J.A. and M.R. Meybodi, 2010. Clustering the wireless Ad Hoc networks: A distributed learning automata approach. J. Parallel Distrib. Comput., 70: 394-405.

28. Webb, A.R., 2002. Statistical Pattern Recognition. 2nd Edn., John Wiley and Sons Ltd., Oxford, UK.

Chapter 9

Performance of Gain Scheduled Generic Model Controller Based on BF-PSO for a Batch Reactor

S. Sujatha and N. Pappa

ABSTRACT

Performance of the gain scheduled Generic Model Controller (GMC) in tracking the batch reactor's optimal temperature profile is considered here. The chosen optimal reactor temperature profile has three switching time interval is obtained by solving optimal control problem off-line by using genetic algorithm. The temperature profile or variation in set point in each time interval is maintained by the proposed controller. The GMC controller parameters are tuned based on the Swarm Intelligence Optimization techniques as Particle Swarm Optimization (PSO) and Bacterial Foraging PSO (BF-PSO) for three switching intervals separately and the integrated performance of gain scheduled GMC has been tested with the batch reactor. The performance of the gain scheduled GMC controller is compared with the single tuned GMC and it provided minimum performance index Integral Square Error (ISE) and Integral Absolute Error (IAE) with improved response for multiple set point tracking.

1. INTRODUCTION

Batch reactors are frequently used in chemical, petro-chemical and bio-chemical industry for the production of various quality products. Processes used are discontinuous and varied. Indeed they are characterized by non-stationary and non-linear system (Baghli and Benyeltou, 2006). The control of a batch reactor consists of the process like filling the reactor with the raw materials; allow it to react by controlling the reactor temperature to meet the required processing condition and shutting down the batch reactor and emptying the reactor. The next batch is started with new filling materials. The heat will be supplied at the starting of the reaction to obtain the desired reaction temperature and then it will be cooled to maintain the proper reaction temperature to get the highest product yield. Production of value added goods in small quantities like medicine in pharmaceutical industries while controlling the polluting waste resources and losses of raw material remains a significant aim of the fine chemical industries. As a result, functioning of the batch reactors at its optimum situation is critical (Aziz et al., 2000).

The best possible control of batch reactors has been recognized as main concentration in the precedent. The primary aim is to find out the optimum reactor temperature profile which yield the utmost desired product and decrease the bi-product as the multi-objective optimization problem. Designing a controller to implement the optimal control profiles or tracking the dynamic set point becomes an important area of research for intrinsically dynamic batch processes in recent years. In an exothermic batch reactor, temperature overshoot is usually avoided because it can cause runaway due to the large amount of heat released at the elevated reactor temperatures. Several authors (Aziz et al., 2000; Mujtaba et al., 2006) have proposed the control system that estimates the amount of heat is being released in the reactor and use it in a feed forward control to compensate the consequence of the heat released.

For solving complex engineering problems, many optimization techniques are emerged by inspiring the social behaviour of swarm of fish, bees and other animals. PSO is a robust stochastic evolutionary computation method developed based on the movement of swarms looking for most fertile feeding location (Hassanzadeh and Mobayen, 2008). Comparatively a new evolutionary computation algorithm named as Bacterial Foraging scheme has also been introduced and received more attraction by the researchers Passino (2002). These optimization algorithms have been applied in various fields like Ad-hoc networking (Sabari and Duraiswamy, 2010), Image processing (Saadi et al., 2010; Ouadfel and Batouche, 2007), Artificial intelligence (Qasem and Shamsuddin, 2010), constrained optimization problem solution (Lu and Chen, 2011) etc.

In the design of optimal control field, to obtain optimal controller parameters, many researchers introduced advanced controllers such as PID auto tuning or self tuning (Ustun, 2007), model based adaptive control, model predictive control (Ansarpanahi et al., 2008), fuzzy control (Ali, 2011), artificial neural network, GA based control (Hassanzadeh et al., 2008; Mosavi, 2011), optimal control, expert control and robust control (Ajlouni and Al-Hamouz, 2004). However, all these controllers required some kind of process model and high-level expertise. It is difficult to develop a general-purpose, user-friendly and smart control system based on these control methods. To keep away from these constraints, the computational intelligence has proposed Particle Swarm Optimization (PSO) and Bacterial Foraging that has opened paths to a new generation of an advanced process control. In recent times, these techniques are mostly preferred for tuning of the PID controller (GirirajKumar et al., 2010; Hassanzadeh and Mobayen, 2008; Korani et al., 2009; Pillay and Govender, 2007; Hooshmand, 2008).

In advanced process control, recently many model based control schemes have been developed. Generic Model Control (GMC) approach has been introduced to design the control action for the non linear process which include the process model itself in the control algorithm structure (Lee and Sullivan, 1988). In this paper, the gain scheduling of GMC controller parameters K1 and K2 for each switching time interval has been implemented by using these Particle Swarm Optimization and Bacterial Foraging techniques.

2. BATCH REACTOR MODEL

A complex reaction scheme is a representative of many industrial reactions (Baghli and Benyeltou, 2006). The batch reactor used by Cott and Macchietto (1989), Aziz et al. (2000) and Sujatha and Pappa (2011) is shown in Fig. 1. Reactions I and II are given by Eq. 1:

$$\begin{aligned} &A + B \rightarrow C \\ &A + C \rightarrow D \quad \text{Reactions I and II} \end{aligned} \tag{1}$$

where, A, B are the raw materials, C is the desired product and D is the waste product.

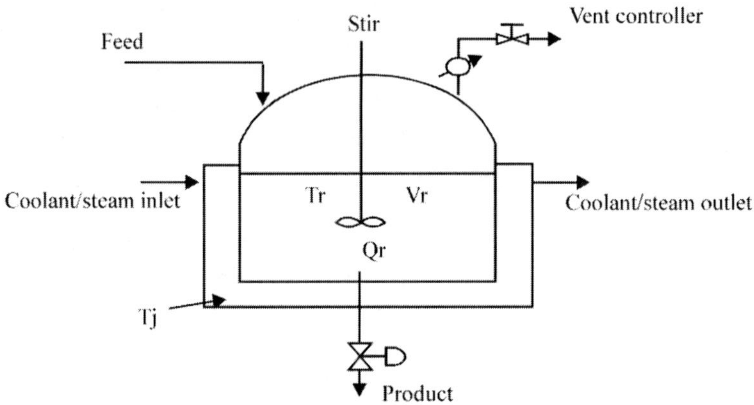

Figure 1: Schematic diagram of a jacketed batch reactor

A batch reactor is inherently dynamic process. Therefore, its model results in a system of differential and algebraic equations. The model equations for the batch reactor can be written as:

$$\frac{dM_A}{dt} = -R_1 - R_2 \qquad (2)$$

$$\frac{dM_B}{dt} = -R_1 \qquad (3)$$

$$\frac{dM_C}{dt} = R_1 - R_2 \qquad (4)$$

$$\frac{dM_D}{dt} = R_2 \qquad (5)$$

$$R_1 = k_1 M_A M_B$$
$$R_2 = k_2 M_A M_C \tag{6}$$

$$k_1 = \exp\left(\frac{k_1^1 - k_1^2}{(Tr + 273.15)}\right) \tag{7}$$

$$k_2 = \exp\left(\frac{k_2^1 - k_2^2}{(Tr + 273.15)}\right) \tag{8}$$

$$\frac{dTr}{dt} = \frac{(Q_r + Q_j)}{M_r C_{pr}} \tag{9}$$

$$\frac{dTj}{dt} = \frac{(T_j^{sp} - T_j)}{\tau_j} - \frac{Q_j}{V_j \rho_j C_{pj}} \tag{10}$$

$$Q_r = -\Delta H_1 R_1 - \Delta H_2 R_2 \tag{11}$$

$$M_r = M_A + M_B + M_c + M_D \tag{12}$$

$$C_{pr} = \frac{C_{pA} M_A + C_{pB} M_B + C_{pC} M_C + C_{pD} M_D}{M_r} \tag{13}$$

$$Q_j = UA(T_j - T_r) \tag{14}$$

where, in the initial values of the above mentioned process parameters are [MA, MB, MC, MD, Tj, Tr] = [12.0, 12.0, 0.0, 0.0, 20.0, 20.0] at t = 0, respectively. The reactor temperature is used as the control variable and is bounded between 20 and 100°C and the jacket temperature is the manipulated variable and it is bounded between 20 to 120°C. The batch time is fixed to 120 min. All the nominal parameters and constant values used in the model equations (Sujatha and Pappa, 2010; Sujatha and Pappa 2011) are given in Table 1.

Table 1 Nominal values of the parameters

Constant	Parameters	Values
C_{pA}	Specific heat capacity of component A	18.0 kcal kmol^{-1}°C
C_{pB}	Specific heat capacity of component B	40.0 kcal kmol^{-1}°C
C_{pC}	Specific heat capacity of component C	52.0 kcal kmol^{-1}°C
C_{pD}	Specific heat capacity of component D	80.0 kcal kmol^{-1}°C
ΔH_1	Heat of reaction of reaction 1	-10000.0 kcal kmol^{-1}
ΔH_2	Heat of reaction of reaction 2	-6000.0 kcal kmol^{-1}
C_p	Mass heat capacity of reactant	0.45 kcal kg^{-1}°C
C_{pj}	Molar heat capacity of component j	0.45 kcal g^{-1}°C
U	Heat transfer coefficient	9.76 kcal min^{-1} m^{2}°C
ρ_j	Density	1000.0 kg^{-1} m^{3}
k_1^1	Pre-exponential rate constant for reaction 1	20.9057
k_1^2	Pre-exponential rate constant for reaction 1	10000
k_2^1	Pre-exponential rate constant for reaction 2	38.9057
k_2^2	Pre-exponential rate constant for reaction 2	17000
V_j	Jacket volume	0.6921 m^{3}
A	Heat transfer area	6.24 m^{2}
Mr	Number of moles of component	1560 kg
τ_j	Jacket time constant	3.0 min

3. OPTIMAL TEMPERATURE PROFILE GENERATION

Genetic Algorithm (GA) is one of the modern optimization methods and the principle is based on the Darwinian Theory of evolutionary process (Mosavi, 2011; Asfaw and Saiedi, 2011). Here, GA is used to obtain the optimal temperature profile in offline. The fitness function for GA is expressed as given the initial charge per unit volume of the reactor, for the fixed batch time optimize the reactor temperature profile so as to maximize the conversion of the desired product and to minimize the waste product is subject to the constraints on the reactor temperature and constraints on the waste product (Sujatha and Pappa, 2011). The obtained optimized temperature profile is [93→92.34→93.57] that yields the maximum product and is shown in Fig. 2.

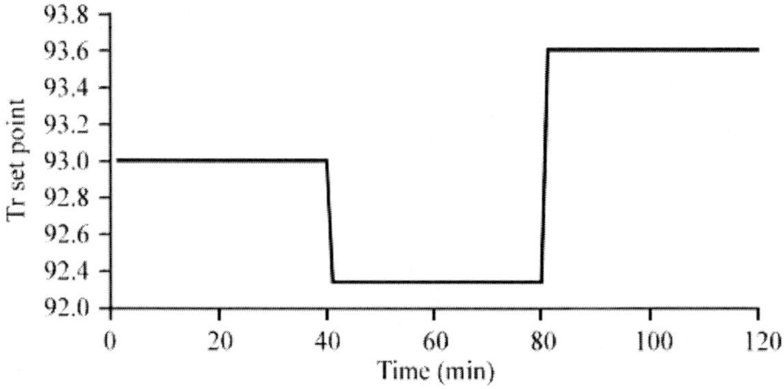

Figure 2: Optimum temperature profile obtained from GA

4. PSO AND BF-PSO TUNED GENERIC MODEL CONTROL (GMC)

In this study, PSO and BF-PSO techniques are used to optimize the GMC controller parameters. The GMC algorithm can be written as:

$$\frac{dx}{dt} = k_1(x_{sp} - x) + k_2\int(x_{sp} - x)dt \tag{15}$$

where, x is the current value and xsp is the desired value of the control variable. The first expression in the algorithm is to bring back the process to steady state due to change in dx/dt. The second expression is introduced to make the zero offset.

For temperature control of the batch reactor, a process model relating the reactor temperature (Tr) to the manipulated variable, the jacket temperature (Tj) is required. Assuming that, the amount of heat retained in the walls of the reactor is small in comparison to the heat transferred in the rest of the system, the following model is given based on the energy balance around the reactor contents:

$$\frac{dT_r}{dt} = \frac{Q_r + UA(T_j - T_r)}{W_r C_{pr}} \quad (16)$$

Replacing x with Tr in the general algorithm the control formulation of GMC is given by the Eq. 17:

$$T_j = T_r + \frac{W_r C_{pr}}{UA}\left[k_1 (T_{rsp} - T_r) + k_2 \int (T_{rsp} - T_r) dt\right] - \frac{Q_r}{UA} \quad (17)$$

where, Tj gives the required jacket temperature trajectory so that the reactor temperature Tr, follows the desired trajectory incorporating the values of GMC tuning parameters k1 and k2. These parameters are tuned by using standard PSO and BF-PSO techniques for these three switching interval [93→92.34→93.57] as shown in Fig. 3.

Particle swarm optimization (PSO): PSO is a computation intelligent technique, which was motivated by the organisms' behaviour such as schooling of fish and flocking of birds. The major advantage of PSO is that it employs the physical movements of the individuals in the swarm and has a flexible and well-balanced mechanism to enhance and to adapt the global and local exploration abilities (Hassanzadeh and Mobayen, 2008).

The global optimizing model is described as follows:

$$\begin{aligned} V_{id}^{t+1} &= \omega V_{id}^{t} + C_1 r_1 (P_{best}^{t} - x_{id}^{t}) + C_2 r_2 (g_{best}^{t} - x_{id}^{t}) \\ x_{id}^{t+1} &= x_{id}^{t} + V_{id}^{t} \end{aligned} \quad (18)$$

where, Vid is the velocity of particle I, representing the distance to be travelled by this particle from its current position, t is the number of iterations, xid represents the particle position, C1 and C2 are the positive constant parameters, r1 and r2 are the random functions in the range [0 , 1], Pbest (local best solution) is the best position of the ith particle, g best (global best solution) is the best position among all particles in the swarm, and ù is the inertia weight which is very important for the convergence behaviour of PSO. A suitable value usually provides a balance between global and local exploration abilities and consequently results in a better optimum solution. Then the following equation is used to adjust to enable quick convergence:

$$\omega = \omega_{max} - (\omega_{max} - \omega_{min})k / k_{max} \tag{19}$$

where, ωmax is the initial weight, ωmin is the final weight, k is the current generation and kmax is the maximum number of generation. Particles' velocities on each dimension are limited to a maximum velocity Vmax.

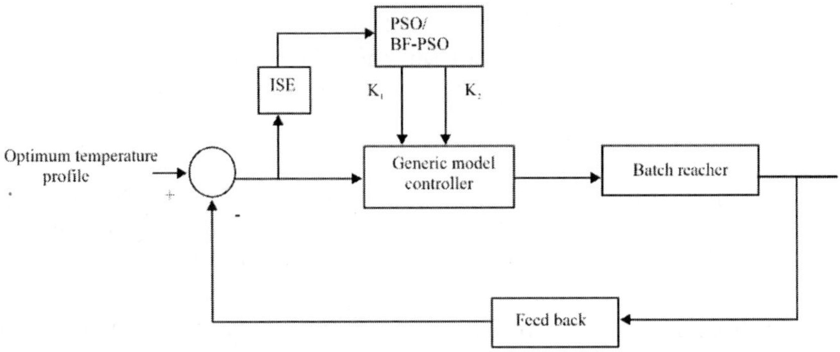

Figure 3: Proposed controller scheme

Bacterial foraging-PSO algorithm (BF-PSO): It is the combination of bacterial foraging and particle swarm optimization algorithm proposed by Korani et al. (2009). In this algorithm, PSO ability is used to exchange social information and BF used in finding a new solution by elimination and dispersal. BF algorithm is based upon search and optimal foraging decision making capabilities of the E. coli bacteria. Each bacterium tries to maximize its obtained energy per each unit of time expended on the foraging process and avoiding noxious substances. It undergoes four stages namely chemotaxis, swarming, reproduction, elimination and dispersal. Chemo-tactic movement is continued until a bacterium goes in the direction of positive nutrient gradient (Shen et al., 2009) (i.e., increasing fitness). The flow chart of BF-PSO has been given in Fig. 4.

5. RESULTS AND DISCUSSION

The number of swarms in the PSO algorithms and number of bacteria in BF-PSO algorithms are varied and the control parameters obtained from these techniques are presented in the Table 2 and 3.

In the PSO tuned GMC, the following values are assigned: dimension is set to two; PSO parameters are set to C1 = 0.12, C2 = 1.2; its momentum has the value of 0.9 and the Integral Square Error (ISE) is used to converge the algorithm.

From Table 2 the variation of number of swarms means the change in number of iterations, while increasing the number of iterations, the PSO settles in an optimized value of GMC controller parameters i.e. K1 and K2 are settled for the three different switching time intervals (0-40, 40-80, 80-120) are shown in Fig. 5.

This optimized GMC tuning parameters are integrated in the controller with each time interval, it is adapted and the simulated response of the batch reactor for this controller setting has been revealed in Fig. 6a and b.

In the BF-PSO based GMC tuning, the following values are assigned:

- Search space dimension is two; number of chemotactic steps are set to ten; length of swim is four; number of reproduction steps are four; number of elimination dispersal event is two; number of bacteria reproduction per generation is half of the number of bacteria; probability of bacteria elimination has a value of 0.25

From the Table 3, the increased number of bacteria makes the iteration changes and the controller values are settled in each time interval as shown in Fig. 7.

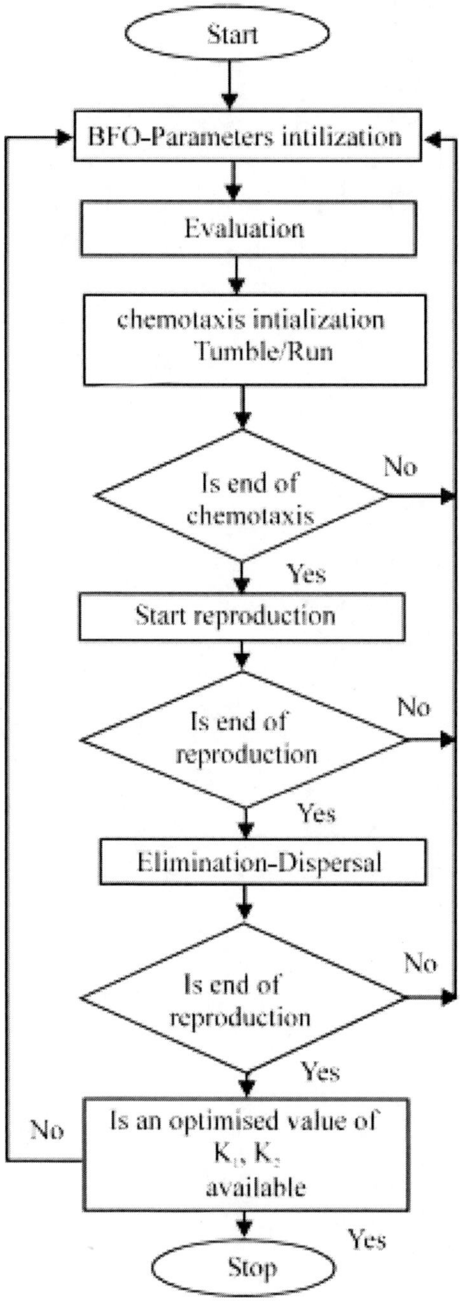

Figure 4: Flow chart for BF-PSO

Table 2: PSO tuned GMC Parameters

	Time interval (0-40) min		Time interval (40-80) min		Time interval (80-120) min	
Number of iterations (Swarms)	K_1	K_2	K_1	K_2	K_1	K_2
10	5.0133	0.3515	4.2243	1.836	6.6776	0.8332
20	5.0605	-0.0505	5.5780	-0.9292	5.9540	2.6751
30	4.7613	1.7015	4.0652	1.3595	5.3578	2.2008
40	4.3062	0.386	5.0528	2.1784	4.689	2.2711
50	4.3062	0.386	5.0661	2.2534	5.1112	2.1578
60	4.1692	0.4156	4.7885	1.8736	4.8489	2.2918
70	4.3696	0.4151	5.0983	2.0976	4.8751	1.8903
80	4.2984	0.3664	4.9563	1.9615	4.9623	2.1338
90	4.3442	0.4397	4.9568	2.0655	4.9244	1.9998
100	4.3151	0.3781	5.0009	2.2092	4.9306	2.0593
120	4.2979	0.3905	5.0074	2.0576	4.9335	2.0379
150	4.3117	0.3993	5.0312	2.0691	4.9367	2.082

Table 3: BF-PSO tuned GMC parameters

Number of iterations (Bacteria)	Time interval (0-40) min		Time interval (40-80) min		Time interval (80-120) min	
	K_1	K_2	K_1	K_2	K_1	K_2
10	4.8514	1.2276	4.8839	1.1144	5.7567	0.6733
20	4.835	1.1109	5.2407	1.035	5.8409	0.911
30	4.8759	1.0963	5.2025	1.7841	5.5816	1.1137
40	4.8242	1.6079	4.8448	1.2100	5.8959	0.8787
50	4.9154	1.3669	5.2882	1.2872	5.3933	1.1326
60	4.8961	0.9508	4.8992	1.1941	5.2351	1.6008
70	4.8240	0.8662	4.7949	0.9699	5.1332	1.8222
80	4.8665	1.1435	4.8499	1.2125	5.2616	1.4794
90	4.9515	1.7238	4.9024	1.1224	5.1705	1.9831
100	4.9423	1.3155	4.9290	1.6254	5.1494	1.9251
120	4.3184	0.4045	4.9666	1.383	5.1656	1.9045
150	4.3627	0.4429	4.8957	1.3258	5.1554	1.9067

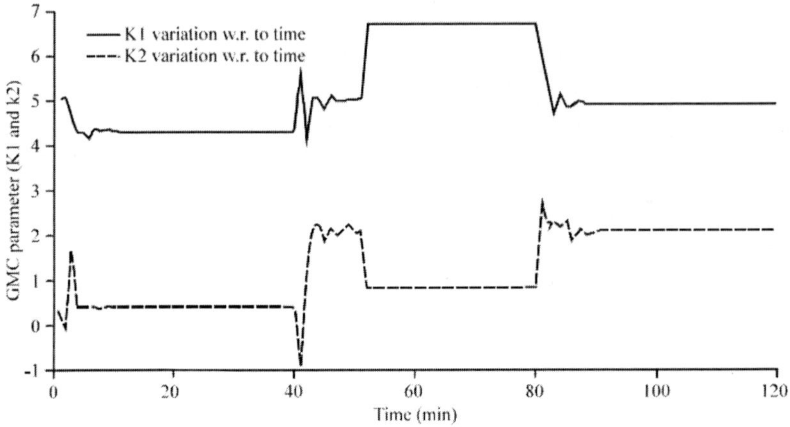

Figure 5: PSO tuned GMC parameters variation with respect to three switching time intervals

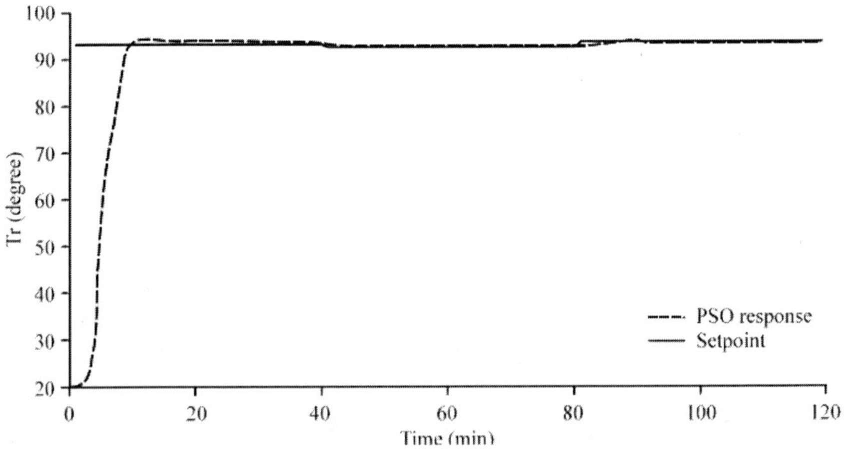

Figure 6a: PSO tuned GMC Response

The response of the integrated BF-PSO tuned GMC for the batch reactor with three switching interval is simulated with the process and shown in Fig. 8a and b.

BF-PSO response produce fast rise time and to reach its set point at time t = 5 min and giving good tracking over the multiple set points than PSO which has rise time t = 10 min as shown in Fig. 9a and b.

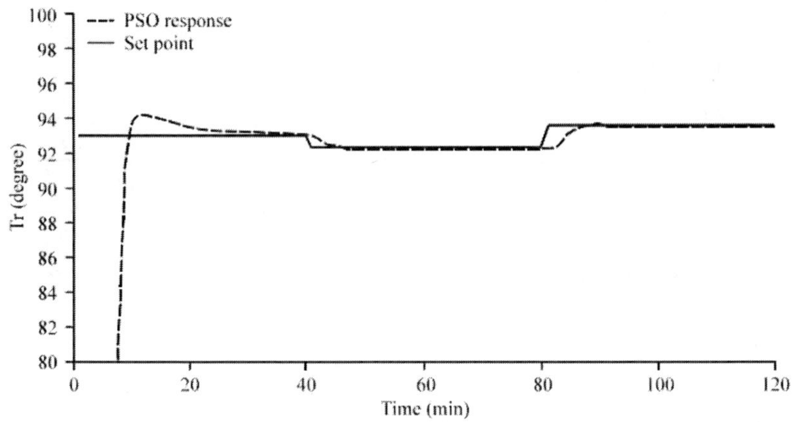

Fig. 6b: PSO tuned GMC Response (Magnified view of the above figure)

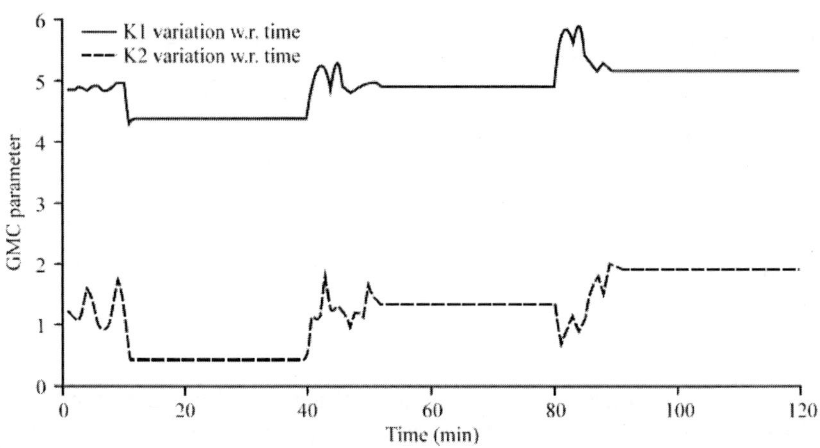

Figure 7: BF-PSO tuned GMC parameters Variation with respect time intervals

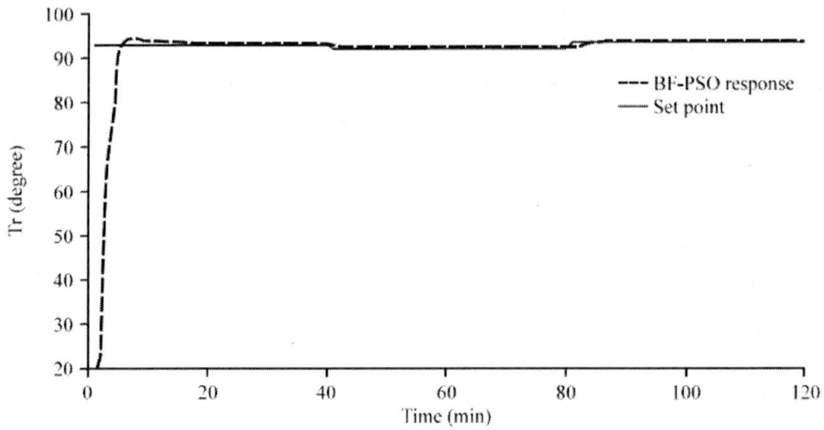

Figure 8a: BF-PSO tuned GMC response

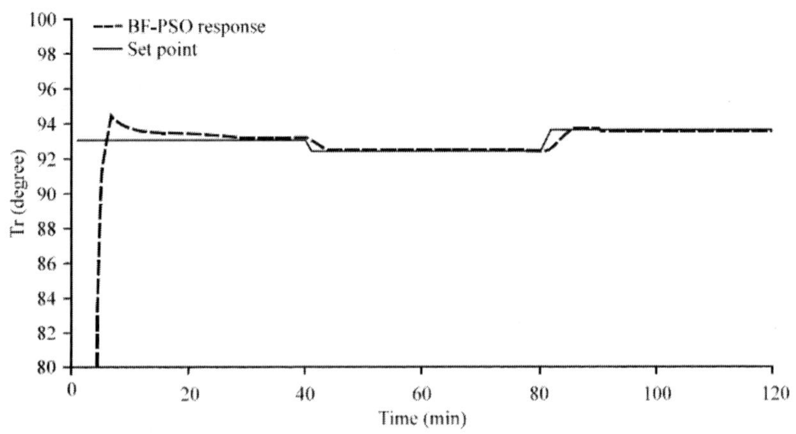

Figure 8b: BF-PSO tuned GMC response (in magnified view)

The proposed controller performance compared with the nominal GMC setting, the single tuned GMC based on PSO and BF-PSO techniques and gain scheduled GMC based on PSO and BF-PSO with its performance measure as Integral Square Error (ISE), Integral Absolute Error (IAE) are presented in the Table 4.

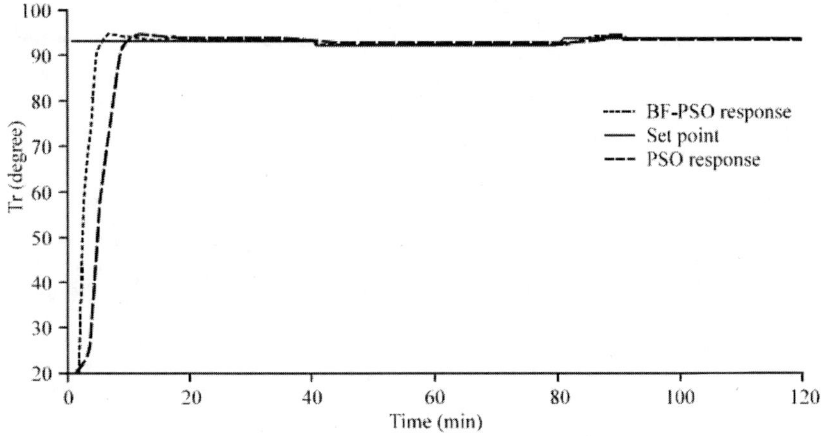

Figure . 9a: Comparison of PSO and BF-PSO tuned GMC response

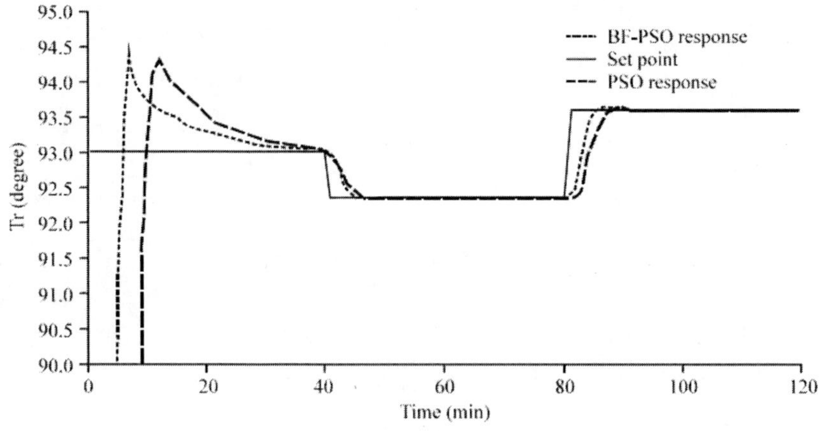

Figure 9b: Comparison of PSO and BF-PSO tuned GMC response (in magnified view)

The performance of gain scheduled GMC based on BF-PSO is better than the other controller responses as shown in Fig. 10a and b and also the performance criteria ISE and IAE are good for the BF-PSO based gain scheduled GMC controller.

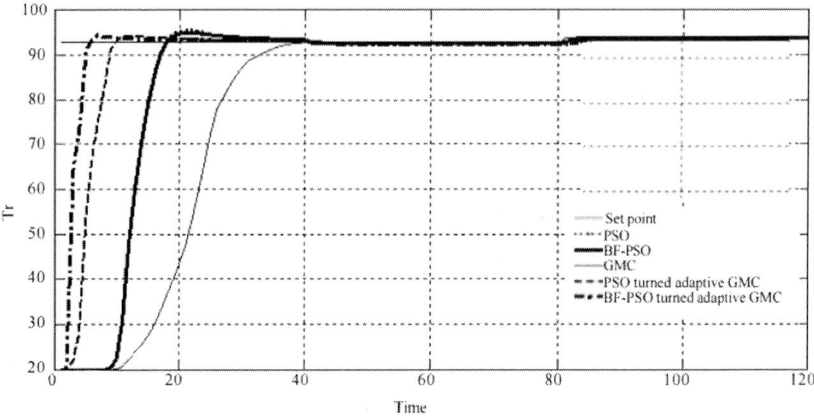

Figure 10a: Comparison of proposed controller with single tuned and common GMC

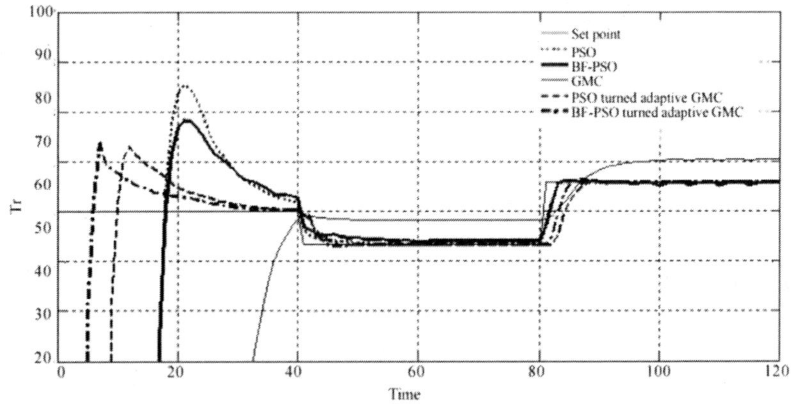

Figure 10b: Comparison of proposed controller with single tuned and common GMC (in magnified view)

Table 4: Controller evaluation

S. No.	Controllers	Optimum settings		Performance criteria	
		K_1	K_2	ISE	IAE
1	GMC	0.12	0.0001	21870	619.3
2	PSO single tuned GMC	4.3393	0.3086	1628	80.93
3	BF-PSO single tuned GMC	4.3762	0.3225	1529	76.75
4	PSO gain scheduled GMC				
	(0-40) min	4.3117	0.3993	612.4	30.067
	(40-80) min	5.0312	2.0691	0.0333	0.1933
	(80-120) min	4.9367	2.082	0.1476	0.4063
5	BF-PSO gain scheduled GMC				
	(0-40) min	4.3627	0.4429	605.3	30.02
	(40-80) min	4.8957	1.3258	0.0332	0.1856
	(80-120) min	5.1554	1.9067	0.1413	0.3956

6. CONCLUSION

Swarm intelligent techniques are popularly used for tuning of PID controller nowadays. These techniques developed from replicating the evolutionary process and the behaviours of biology with basic mathematical operations. Therefore, they are simple and easy to be implemented. And also they don't insist to meet the state of differentiability, convexity and other conditions for mathematical description of the problem. Here, PSO and BF-PSO tuning has been attempted to tune Generic Model controller (GMC) parameters for tracking the optimal reactor temperature profile to the batch reactor. The simulation results presented shows the betterness of the hybrid optimization method validated which replicates the efficiency of the BF-PSO in terms of time domain specifications. It is also observed that BF-PSO based gain scheduled GMC i.e., as per the optimal temperature profile the set point is varying with three switching time interval; so the GMC parameters are also optimized in these three switching interval, and the integrated response of this gain scheduled adaptive GMC has produced minimum performance index ISE and IAE with improved responses for multiple set point tracking and also in terms of time domain specifications. Further, the work can be applied for the complex process also by including disturbances.

REFERENCES

1. Ajlouni, N. and S. Al-Hamouz, 2004. Genetic design of neural PID plus feed forward controllers. Inform. Technol. J., 3: 6-11.

2. Ali, H.K., 2011. Fuzzy controller design of servo system. Asian J. Applied Sci., 4: 403-413.

3. Ansarpanahi, S., S.B.M. Noor and M.H. Marhaban, 2008. Stability study of model predictive control in presence of undesirable factors. J. Applied Sci., 8: 3683-3689.

4. Asfaw, T.D. and S. Saiedi, 2011. Optimal short-term cascade reservoirs operation using genetic algorithm. Asian J. Applied Sci., 4: 297-305.

5. Aziz, N., M.A. Hussain and I.M. Mujtaba, 2000. Performance of different types of controllers in tracking optimal temperature profiles in batch reactors. Comput. Chem. Eng., 24: 1069-1075.

6. Baghli, H. and M. Benyettou, 2006. Modeling and simulation of thermal transfer in batch reactor. J. Applied Sci., 6: 1327-1331.

7. Cott, B.J. and S. Macchietto, 1989. Temperature control of exothermic batch reactors using generic model control. Ind. Eng. Chem. Res., 28: 1177-1184.

8. GirirajKumar, S.M., D. Jayaraj and Anoop. R. Kishan, 2010. PSO based tuning of a PID controller for a High performance drilling machine. Int. J. Comp. Appl., 1: 12-18.

9. Hassanzadeh, I. and S. Mobayen, 2008. PSO based controller design for rotary inverted pendulum system. J. Applied Sci., 8: 2907-2912.

10. Hassanzadeh, I., S. Mobayen and G. Sedaghat, 2008. Design and implementation of a controller for magnetic levitation system ssing genetic algorithms. J. Applied Sci., 8: 4644-4649.

11. Hooshmand, R.A., 2008. Optimal design of load shedding and generation reallocation in power systems using fuzzy particle swarm optimization algorithm. J. Applied Sci., 8: 2788-2800.

12. Korani, W.M., H.T. Dorrah and H.M. Emara, 2009. Bacterial foraging oriented by particle swarm optimization strategy for PID tuning. Proceedings of the IEEE International Symposium on Computational Intelligence in Robotics and Automation, December 15-18, 2009, Daejeon, South Korea, pp: 445-450.

13. Lee, P.L. and G.R. Sullivan, 1988. Generic model control (GMC). Comput. Chem. Eng., 12: 573-580.

14. Lu, H. and X. Chen, 2011. A new particle swarm optimization with a dynamic inertia weight for solving constrained optimization problems. Inform. Technol. J., 10: 1536-1544.

15. Mosavi, M.R., 2011. Applying genetic algorithm to fast and precise selection of GPS satellites. Asian J. Applied Sci., 4: 229-237.

16. Mujtaba, I.M., N. Aziz and M.A. Hussain, 2006. Neural network based modeling and control in batch reactor. Chem. Eng. Res. Des., 84: 635-644.

17. Ouadfel, S. and M. Batouche, 2007. AntClust: An ant algorithm for swarm-based image clustering. Inform. Technol. J., 6: 196-201.

18. Passino, K.M., 2002. Biomimicry of bacterial foraging for distributed optimization and control. IEEE Control Syst., 22: 52-67.

19. Pillay, N. and P. Govender, 2007. A particle swarm optimization approach for model independent tuning of PID control loop. Proceedings of the IEEE Meeting on AFRICON, September 26-28, 2007, Windhoek, Namibia, pp: 1-7.

20. Qasem, S.N. and S.M. Shamsuddin, 2010. Generalization improvement of radial basis function network based on multi-objective particle swarm optimization. J. Artif. Intell., 3: 1-16.

21. Saadi, S., M. Bettayeb and A. Guessoum, 2010. Optimal approach for neutron images restoration using particle swarm optimization algorithm with regularization. J. Applied Sciences, 10: 517-525.

22. Sabari, A. and K. Duraiswamy, 2010. Swarm based reliable and power efficient multicast routing in mobile Ad Hoc network. Inform. Technol. J., 9: 1383-1389.

23. Shen, H., Y. Zhu, X. Zhou, H. Guo and C. Chang, 2009. Bacterial foraging optimization algorithm with particle swarm optimization strategy for global numerical optimization. Proceedings of the first ACM/SIGEVO Summit on Genetic and Evolutionary Computation, June 12-14, 2009, Shanghai, China, pp: 497-504.

24. Sujatha, S. and N. Pappa, 2010. GMC design based on SVM model of batch reactor. Int. J. Comput. Appl., 1: 76-80.

25. Sujatha, S. and N. Pappa, 2011. Realization of non linear controllers in batch reactor using GA and SVM. Int. J. Chem. Reactor Eng., 9: A67-A67.

26. Ustun, S.V., 2007. GA-based optimization of PI speed controller coefficients for ANN-modelled vector controlled induction motor. J. Applied Sci., 7: 4001-4006.

Chapter 10

Decentralized self-selection of swarm trajectories: from dynamical systems theory to robotic implementation

ABSTRACT

A system with parameter variation within bound creates interval in coefficients in the system polynomial and hence it is called as interval system. In this study, a model order reduction for linear interval system is attempted. A convergence technique is attempted to reduce the numerator polynomial. The denominator polynomial is reduced by using the methods such as retention of Markov parameter and time moments technique. A numerical approach is proposed to compute the reduced order interval model for a higher order linear time variant model. Also in this study, a novel technique using Particle Swarm assisted Bacterial Foraging Optimization (PSO-BFO) based hybrid algorithm is proposed to search the PID controller parameters such as Kp, Ki and Kd. The algorithm is to obtain the best possible PID parameters with Integral Squared Error (ISE) criterion minimization is as the objective function. Initially the controller parameters are obtained for reduced order model; then it is tested with the higher order model. A simulation is carried out to shows that the effectiveness of proposed model reduction algorithm and the controller performance. From the result it is observed that, the projected method provides enhanced performance for the interval system.

1. INTRODUCTION

In Industries and academics, model based analysis are widely proposed to study the transient and steady state behaviour of the physical systems such as machineries, process loops, etc. Most of the modeling methods proposed in the literature may provide only the approximated models around the operating region of the process under study. The conventional modeling methods may fail to provide the approximated model when the system has parametric perturbation. Since, in this study, we proposed an interval system analysis accounting the possible perturbations. A variety of modeling methods (Dolgin and Zeheb, 2004) has been discussed to construct an approximated lower order model for the real time systems.

Model based analysis is used to study the behaviour of the physical system. Many attempts have been made to model the physical system with parametric uncertainties as interval system (Dolgin and Zeheb, 2004). When the real time system is modeled, its order is too large. Hence, analysis like steady state and transient response is very difficult. So the model reduction is very much essential for the analysis and control of physical system. Some of the model reduction techniques have been proposed by various researchers for interval system (Younseok, 2007). Model order reduction of interval system without sacrificing the response of higher order model is a challenging task. But the validity of the method is based on the resulting error obtained by the model reduction (Babu and Pappa, 2008).

Identification of fault is an important task in intelligent control system, since a fault can lead to reduction in performance even breakdown. Fault detection identifies the deviation of set point in the monitored system. Ideal system and the model behave exactly the same, but in real time system fault is detected when the behaviour of the system differs from reference beyond the tolerance level.

The method presented by Lepschy and Viaro (1983), is a mixed method using differentiation and Pade's reduction techniques which avoids the use of gain factor. In Lucas model reduction technique (Lucas, 1988, 1992) for an Interval system is based on differentiation technique for both the numerator and denominator. The method has a serious limitation, since it has the formation of two Routh arrays for the numerator and denominator, it also always retains only one time moment of the original system which affects the time response of the system. This method has a one more serious limitation; it cannot be implemented for systems having a difference of more than one in the order of denominator and numerator polynomials.

In this study, a convergence technique is attempted to reduce denominator polynomial and the numerator polynomial is reduced using biased model reduction which helps to retain the first (k-1) time moments and Markov parameter of the original system (Hsu et al., 2007). Using Kharitonov polynomial, reduced order model is converted into 16 set of polynomials of transfer function model that describes the system behaviour with in bound.

In recent years, nature inspired algorithms, such as Particle Swarm Optimization (PSO) and Bacterial Foraging Algorithm (BFO) methods are widely proposed to find solutions for highly complex engineering optimization problems.

PSO is proposed by Kennedy and Eberhart (1995), it is a swarm based optimization technique inspired by the flock of birds and school of fish. The basic and the modified form of this algorithm was widely proposed by the researchers to find the best possible solutions for their problems. BFO is also one of the nature inspired algorithm which mimics the foraging methods of E. coli bacterium. This algorithm was widely proposed for parameter optimization. Rajinikanth and Latha (2011) proposed the BFO algorithm to optimize the PID controller parameter for a class of unstable system. With a comparative study they concluded that even though it provides the best optimized value, its speed of convergence is slower than the PSO algorithm.

The Integral Squared Error (ISE) criterion is used to guide the Bacterial Foraging based Particle Swarm Optimization (BF-PSO) algorithm to search the controller parameters such as Kp, Ki and Kd. It is used to find a set of controller parameters are obtained for each Kharitonov polynomial. A hull set for each controller parameter is calculated from the 16 set of controller parameters which are obtained for 16 Kharitonov polynomial using Model Interval Analysis (MIA). Controller performance is analyzed by computing ISE and compared withBandyopadhyay et al. (1997) method. Using the generalized Kharitonov theorem and Kharitonov rectangle, stability of original and reduced order interval model is verified.

In control literature, recently soft computing approach based PID tuning methods are widely proposed by the researchers for a class of stable and unstable system models. Ali and Majhi (2006) and Rajinikanth and Latha (2011) discussed PSO based PID tuning technique for stable and unstable system. Korani et al. (2009), proposed a hybrid PSO-BFO algorithm to optimize the controller parameter for stable systems and compared the performance with the PSO and BFO based methods. Since, in this study, PSO-BFO based hybrid optimization algorithms is attempted. In this method, the PSO algorithm is used to speed up the process of BFO in order to achieve faster convergence with best optimized parameter.

2. INTERVAL SYSTEM ANALYSIS

Recent trend in the area of system modeling is to model the system with uncertainties. Presence of uncertainties in the system components, process input, disturbances, measurement noise, leakage in joints, system model will vary frequently. One way to model the system with uncertainty is interval system. In interval system the parameter variations within bounds creates intervals in the coefficients of the system polynomial. i.e., a model which has the parameter with interval coefficients like s2+[5,6]s+[0.23,0.34]. Each coefficient will have a lower and upper limit. Interval analysis is done based on the interval arithmetic equation (1-4). Stability of Interval system can be analyzed using Kharitonov stability criterion and Kharitonov rectangle (Tan and Atherton, 2000).

2.1 Interval arithmetic: A real interval x is a nonempty set of real members:

where, is called the infimum and is called the supremum. The set of all intervals over R is denoted by IR. Let x, y be two real interval numbers. The arithmetic operations like addition, subtraction, multiplication and division can be performed with the following equations:

Addition: $$x+y = [\underline{x} + \underline{y}, \overline{x} + \overline{y}] \quad (1)$$

Subtraction: $$x-y = [\underline{x} + \overline{y}, \overline{x} + \underline{y}] \quad (2)$$

Multiplication: $$x \times y = [\min\{\underline{x}y + \underline{x}\overline{y}, \overline{xy}, \overline{yx}\}, \max\{\underline{x}y, \underline{x}\overline{y}, \overline{x}\underline{y}, \overline{yx}\}] \quad (3)$$

Division:

$$\frac{x}{y} = \begin{cases} [\overline{x}/\underline{y}, \infty] & \text{if } \overline{x} \le 0 \text{ and } \overline{y} = 0, \\ [-\infty \overline{x}/\underline{y}] \mathrm{U} [\overline{x}/\underline{y}, \infty] & \text{if } \overline{x} \le 0 \text{ and } \underline{y} < 0 < \overline{y}, \\ [-\infty \overline{x}/\overline{y}] & \text{if } \overline{x} \le 0 \text{ and } \underline{y} = 0, \\ [-\infty, \infty] & \text{if } \underline{x} < 0 \overline{x}, \\ [-\infty \underline{x}/\underline{y}] & \text{if } \overline{x} \le 0 \text{ and } \overline{y} = 0, \\ [-\infty \underline{x}/y] \mathrm{U} [\underline{x}/\underline{y}, \infty] & \text{if } \underline{x} \ge 0 \text{ and } \underline{y} < 0 < \overline{y}, \\ [\underline{x}/\underline{y}, \infty] & \text{if } \underline{x} \ge 0 \text{ and } \underline{y} = 0, \end{cases} \tag{4}$$

3. GENERALIZED KHARITONOV THEOREM

Consider the numerator polynomial of the open loop system is:

$$GP(s) = [\underline{b}_{n-1}, \overline{b}_{n-1}]S^{n-1} + [\underline{b}_{n-2}, \overline{b}_{n-2}]S^{n-2} + \ldots + [\underline{b}_1, \overline{b}_1]S + [\underline{b}_0, \overline{b}_0] \tag{5}$$

The numerator polynomial N(s) is split into two components g(s) and h(s),

Where:

g(s)=Polynomial of even degree

h(s)=Polynomial of odd degree

Defining the two even polynomials:

$$\begin{aligned} G_1(S) &= \underline{b}_0 + \overline{b}_2 s^2 + \underline{b}_4 s^4 + \cdots \\ G_2(S) &= \underline{b}_0 + \overline{b}_2 s^2 + \underline{b}_4 s^4 + \cdots \end{aligned} \tag{6}$$

Defining the two odd polynomials:

$$\begin{aligned} h_1(S) &= \underline{b}_1 s + \overline{b}_3 s^3 + \underline{b}_5 s^5 + \cdots \\ h_2(S) &= \overline{b}_1 s + \overline{b}_3 s^3 + \overline{b}_5 S^5 + \cdots \end{aligned} \tag{7}$$

Then Kharitonov Polynomials can be formulated by Eq. 8:

$$
\begin{aligned}
K_{11}N(S) &= g_1(s) + h_1(s) \\
K_{12}N(S) &= g_1(s) + h_2(s) \\
K_{21}N(S) &= g_2(s) + h_1(s) \\
K_{22}N(S) &= g_2(s) + h_2(s)
\end{aligned} \tag{8}
$$

The Kharitonov polynomials [K11N(s), K12N(s), K21N(s), K22N(s)] and [K11D(s), K12D(s), K21D(s), K22D(s)] are obtained using Eq. 6-8 for the numerator and denominator, respectively. Kharitonov theorem states that, the robust stability of the family of the closed-loop interval polynomials is guaranteed if and only if all Kharitonov polynomials of the system are stable. Obviously, consists of 32 segment polynomials which are generally referred to as generalized Kharitonov segment polynomials. It follows that the entire family of the closed-loop polynomials D(s,p,q,a,b) is Hurwitz stable if and only if the 32 generalized Kharitonov segment polynomials DE(s, l) are all Hurwitz stable.

4. HYBRID OPTIMIZATION ALGORITHM

Particle swarm optimization (PSO) algorithm: Particle Swarm Optimization (PSO) (Kennedy and Eberhart, 1995; Ren and Zhong, 2011) algorithm is a swarm intelligence technique, attempts to mimic the natural way of group communication of individual knowledge. It is a population based search algorithm where each individual is referred to as particle. In this method, a population of swarm is initialized with random positions Si and velocities Vi. At the beginning, each particle of the population is scattered randomly throughout the entire search space and with the guidance of the performance criterion, the flying particles dynamically adjust their velocities according to their own flying experience and their companions flying experience. Each particle remembers its best position obtained so far which is denoted pbest and the overall best out of all the particles in the population is called gbest:

$$v_i(t+1) = w^t v_i(t) + C_1 * r_1(pbest - S_i(t)) + C_2 * r_2(gbest - S_i(y)) \tag{9}$$

$$W^t = (W_{max} - Iter)x\left[\frac{W_{max} - W_{min}}{Iter_{max}}\right] \tag{10}$$

Bacterial foraging optimization (BFO) algorithm: In the process of foraging behaviour of E. coli bacteria undergo four stages, namely, chemotaxis, swarming, reproduction and elimination and dispersal (Passino, 2002). In search space, BFO (Supriyono and Tokhi, 2011) seek optimum value through the chemotaxis of bacteria and realize quorum sensing via assemble function between bacterial and satisfy the evolution rule of the survival of the fittest make use of reproduction operation and use elimination-dispersal mechanism to avoiding falling into premature convergence.

Chemotaxis: The motion patterns that the bacteria will generate in the presence of chemical attractants and repellents are called chemotaxis. For E. coli, this process was simulated by two different moving ways: run or tumble.

Swarming: An interesting group behavior has been observed for several motile species of bacteria including E. coli and S. typhimurium. When a group of E. coli cells is placed in the center of a semisolid agar with a single nutrient chemo-effector, they move out from the center in a traveling ring of cells by moving up the nutrient gradient created by consumption of the nutrient by the group.

Reproduction: According to the rules of evolution, individual will reproduce themselves in appropriate conditions in a certain way. For bacterial, a reproduction step takes place after all chemotactic steps.

Elimination-dispersal: In the evolutionary process, elimination and dispersal events can occur such that bacteria in a region are killed or a group is dispersed into a new part of the environment due to some influence. They have the effect of possibly destroying chemotactic progress, but they also have the effect of assisting in chemotaxis, since dispersal may place bacteria near good food sources (Fig. 1).

5. CONTROLLER TUNING

The following flowchart explains the controller tuning method using PSO.

Proposed algorithm is used to find the controller parameter such as Kp, Ki and Kd for all the 16 set of Kharitonov polynomials of the reduced order model. BF-PSO algorithm is performed with the following parameter Size of the swarm is 10, Dimension of the problem is 3 and c1 = 0.12; c2 = 1.2 and momentum of inertia (w = 0.9). PID controller is designed for all the set of

Kharitonov polynomials. Finally hull set of controller parameter is identified as interval controller gain.

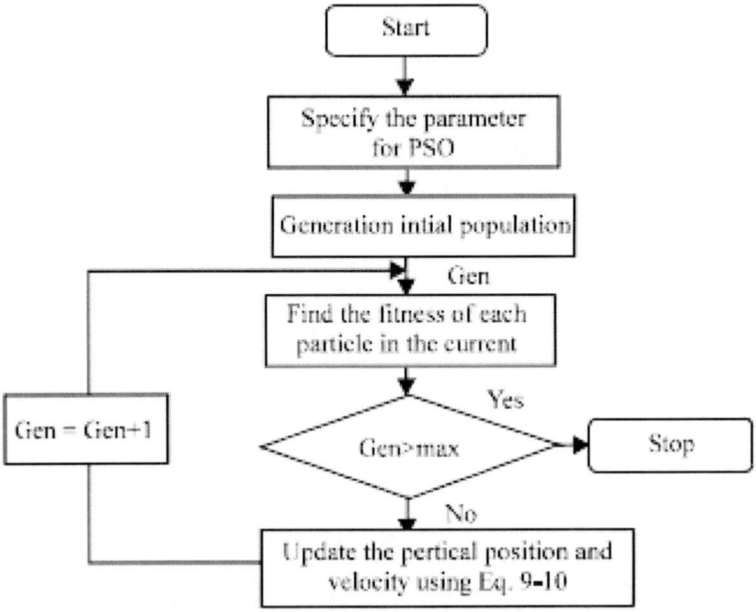

Figure 1: Flowchart of PSO for tuning of controller parameter

6. PROPOSED METHOD FOR MODEL REDUCTION:

A typical nth order interval system is represented by its transfer function:

$$G_n(S) = \frac{(\underline{B}_{n-1}, \overline{B}_{n-1})S^{n-1} + (\underline{B}_{n-2}, \overline{B}_{n-2})S^{n-2} + \cdots (\underline{B}_1, \overline{B}_1)S + (\underline{B}_0, \overline{B}_0)}{(\underline{A}_n, \overline{A}_n)S^n + (\underline{A}_{n-1}, \overline{A}_{n-1})S^{n-1} + \cdots (\underline{A}_1, \overline{A}_1)S + (\underline{A}_0, \overline{A}_0)} \quad (11)$$

Then the reduced (kth) order model:

$$G_k(S) = \frac{(\underline{b}_{k-1}, \overline{b}_{k-1})S^{k-1} + (\underline{b}_{k-2}, \overline{b}_{k-2})S^{k-2} + \cdots (\underline{b}_1, \overline{b}_1)S + (\underline{b}_0, \overline{b}_0)}{(\underline{a}_k, \overline{a}_k)S^n + (\underline{a}_{k-1}, \overline{a}_{k-1})S^{k-1} + \cdots (\underline{a}_1, \overline{a}_1)S + (\underline{a}_0, \overline{a}_0)}$$

$$G_k(S) = \frac{N_k(S)}{D_k(S)} \quad (12)$$

Reduced order denominator: Where is a denominator polynomial coefficient, having interval structure. The kth order reduced model denominator polynomials Dk(s) where (k = 1,2,3….n-1) are obtained by the new algorithms suggested in this study. This method is based on the truncation technique.

For k = 1:

$$D_1(s) = (\underline{A}_0, \overline{A}_0) + \frac{(n-1)C_{n-1}}{n-1}(\underline{A}_1, \overline{A}_1) \tag{13}$$

For k = 2:

$$D_2(s) = (\underline{A}_0, \overline{A}_0) + \frac{(n-1)C_{n-2}}{n-1}(\underline{A}_1, \overline{A}_1)s + \frac{(n-2)C_{n-2}}{nc_{n-2}}(\underline{A}_1, \overline{A}_1)s^2 \tag{14}$$

General representation of the kth order denominator is:

$$D_k(s) = \sum_{i=1}^{k+1}(\underline{A}_{i-1}, \overline{A}_{i-1})\frac{(n-1+1)C_{n-k}}{nC_{n-k}}s^{i-1} \text{ where, } k = 1, 2, 3, \ldots \tag{15}$$

Reduced order numerator: The numerator of reduced order Nk(s) polynomial can be obtained by the following method:

$$N_k(s) = N_{kt}(s) + N_{km}(s)$$

where, Nkt (s) is Numerator term from time constants, Nkm (s) is Numerator term from Markov parameter, k = t+m:

$$= T_1 + T_2 s + \ldots + T_t s^{k-m+1} + M_m s^{k-m} \ldots + M_2 s^{k-2} + M_1 s^{k-1} \tag{16}$$

In general:

$$T_t(s) = (\underline{a}_0, \overline{a}_0)\frac{(\underline{B}_{t-1}, \overline{B}_{t-1})}{(\underline{A}_0, \overline{A}_0)} \tag{17}$$

$$[\underline{M}_i, \overline{M}_i] = \frac{1}{(\underline{An}, \overline{An})}\sum_{j=0}^{m}(\underline{B}_{(n-1)}, \overline{B}_{(n-1)})(\underline{a}_{(k-(m-i))}, \overline{a}_{(k-(m-i))}) - \sum_{j=0}^{m-1}(\underline{Mj}, \overline{Mj})[\underline{A}_{(n-(m-i))}, \overline{A}_{(n-(m-i))}] \tag{18}$$

With Markov parameter .

7. ILLUSTRATIONS

Let the seventh order interval system obtained from the literature:

$$G(s) = \frac{[1.9,\ 2.1]S^6 + [24.7,\ 27.3]S^5 + [157.7,\ 174.3]S^4 + [541.9,\ 599.02]S^3 + [929.9,\ 1027.8]S^2 + [721.81,\ 797.79]S + [187.055,\ 206.745]}{[.95,\ 1.05]S^7 + [8.779,\ 9.703]S^6 + [52.2,\ 57.8]S^5 + [182.8,\ 202.1]S^4 + [429.02,\ 474.1]S^3 + [572.4,\ 632.7]S^2 + [325.2,\ 359.5]S + [57.35,\ 63.389]} \quad (19)$$

A second order model is obtained using the suggested method, retaining initial time moment and one Markov parameter of the given system i.e., k = 2, m = 1, t = 1.

The second order model denominator can be obtained using Eq. 14:

$$= [27.26,\ 30.13]s^2 + [92.937,\ 102.72]s + [57.352,\ 63.789] \quad (20)$$

The numerator N2(s) to retain initial Markov parameter and time moment is obtained using Eq. 17 and 18:

$$(\underline{T}_1,\ \overline{T}_1 = (169.24,\ 229.64) \quad (21)$$

Markov parameter can be calculated by:

$$[M_1,\ M_1] = \frac{1}{[\underline{A_4,\ A_4}]}\,(\underline{B}_3,\ \overline{B}_3)\ (\underline{a}_{3-(1-1)},\ \bar{a}_{3-(1-1)}) - 0*(\underline{A}_3,\ \overline{A}_3) \quad (22)$$

$$(M_1,\ M_1) = [49.32,\ 66.6]$$

Hence, the reduced 2nd order numerator can be obtained using Eq. 21 and 22:

$$N_2(S) = [49.32,66.6]S + [169.24,229.64] \quad (23)$$

Hence, the 2nd order reduced order system using Eq. 20 and 23:

$$G_2(S) = \frac{[49.32,66.6]s + [169.24,229.64]}{[27.26,30.13]s^2 + [92.937,102.72]s + [57.352,63.789]} \quad (24)$$

Kharitonov polynomials of numerator polynomial of Eq. 19 are:

$$K_{u1} = 1.9s^6+24.7s^5+174.3s^4+599.02s^3+929.96s^2+721.81s+206.75$$
$$K_{u1} = 1.9s^6+24.7s^5+174.3s^4+599.02s^3+929.96s^2+721.81s+206.75$$
$$K_{u2} = 2.1s^6+27.3s^5+157.7s^4+541.98s^3+1027.8s^2+797.79s+187.06$$
$$K_{u3} = 2.1s^6+24.7s^5+157.7s^4+599.02s^3+1027.8s^2+721.81s+187.06$$
$$K_{u4} = 1.9s^6+27.3s^5+174.3s^4+541.98s^3+929.96s^2+797.79s+206.75 \quad (25)$$

Kharitonov polynomials of denominator polynomial of Eq. 19 are:

$$K_{d1} = 0.95s^7+8.779s^6+57.8s^5+202.1s^4+429.02s^3+572.4s^2+359.5s+63.389$$
$$K_{d2} = 1.05s^7+9.703s^6+52.2s^5+182.8s^4+474.1s^3+632.7s^2+325.2s+57.35$$
$$K_{d3} = 1.05s^7+8.779s^6+52.2s^5+202.1s^4+474.1s^3+572.4s^2+325.2s+63.389$$
$$K_{d4} = 0.95s^7+9.703s^6+57.28s^5+182.8s^4+429.02s^3+632.7s^2+359.5s+57.35 \quad (26)$$

Using Eq. 25-26 Kharitonov rectangle was drawn and analysis the stability of higher order model.

The characteristic equation of the closed loop system with unity feedback is Eq. 27.

$$1+GH(s) = [.95,\ 1.05]S^7 + [10.967,\ 11.8]S^6 + [76.9,\ 85.1]S^5 + [340.5,\ 376.4]S^4$$
$$+[970.92, 1073.12]S^3 + [1502.3, 1660.5]\ S^2 + [1046.9, 1157.29]S + [244.4,\ 270.1] \quad (27)$$

Characteristic equation of Second order reduced model is:

$$1+G_rH(s) = [27.26, 30.13]s^2+[4.72, 6.21]s+[7.52, 10.764] \quad (28)$$

Using interval analysis reduced order model is modified as:

$$1+G_rH(s) = [1,1]s^2+[4.72, 6.21]s+[7.52, 10.764] \quad (29)$$

8. RESULTS AND DISCUSSION

In this study, a higher order model is obtained from the literature (Tomar et al., 2009), a seventh order stable model is reduced into second order stable model by using convergence technique. Stability of both higher order model and reduced order model can be analyzed using Kharitonov rectangle. In general Kharitonov rectangle is obtained between edge polynomial verses frequency. Figure 2a-c represents the Kharitonov rectangle of stable higher order model. It passes through seven quadrants; hence order of the system is seven. Also it encircles the origin in anticlockwise direction without passing through the origin, so the system is stable system.

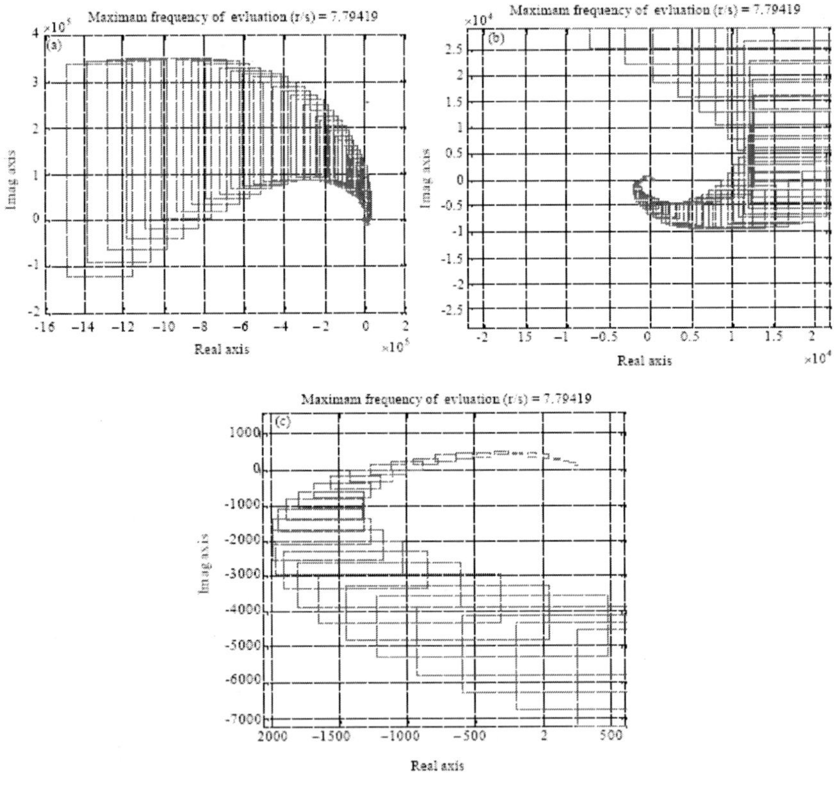

Figure 2(a-c): (a) Kharitonov rectangle for higher order system, (b) Enlarged view of (a) and (c) More enlarged view of (a)

Figure 3 shows the Kharitonov rectangle of stable reduced order model. It has been drawn between the real axis and imaginary axis of Kharitonov edge polynomials obtained for various frequencies. Figure 3 passes two quadrants (first and second) without passing through origin, hence the reduced order model is also stable and order of the model is 2.

It passes through two quadrants and encircles the origin with anticlockwise direction. Anticlockwise encirclement of origin shows stability of the system. To study the steady state and transient state behaviour of the system, step response and impulse response has been obtained for both higher and reduced order model. Figure 4 shows the Step response of the Kharitonov polynomial of the higher and reduced order models. Among Four responses curve, two is drawn for higher order model, remaining two is for reduced order model. Reduced order responses are within higher order response.

To study the system with sudden disturbance, impulse response is used. Figure 5 shows the impulse response of the Kharitonov polynomial of the higher and reduced order models. Similar to step response analysis, impulse response has been obtained for both higher and reduced order model. Here also reduced order model response curve are within the higher order model response curve.

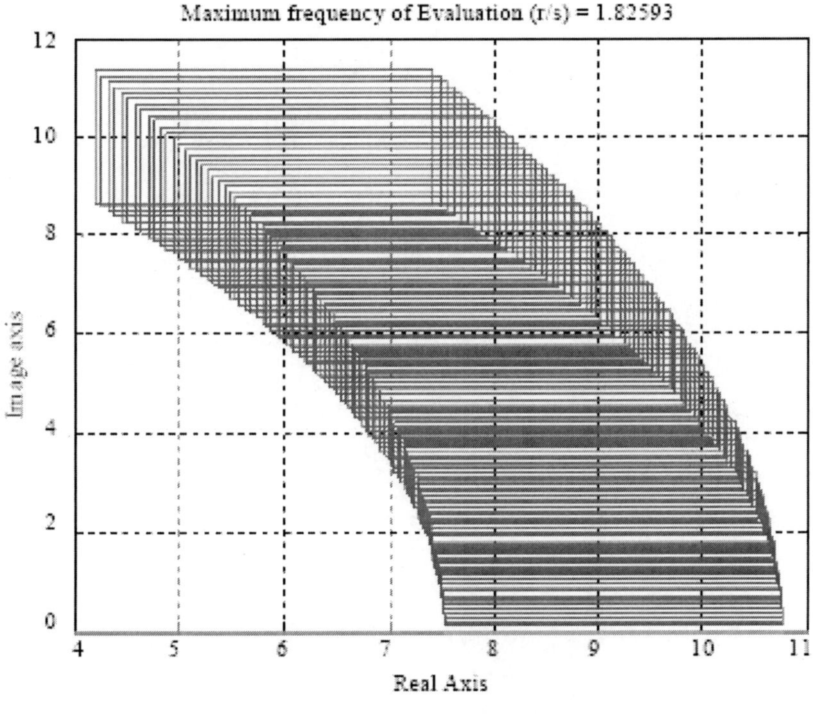

Figure 3: Kharitonov rectangle for reduced order model

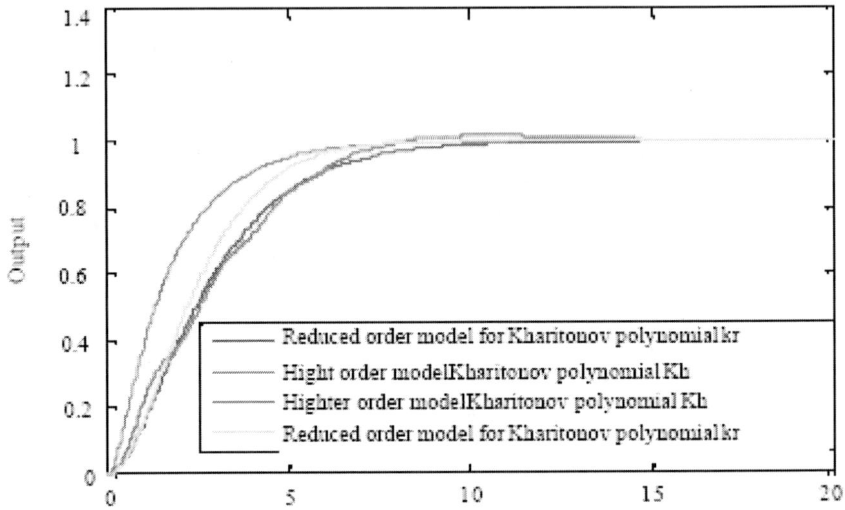

Figure 4: Comparison of step response of higher order and reduced order model

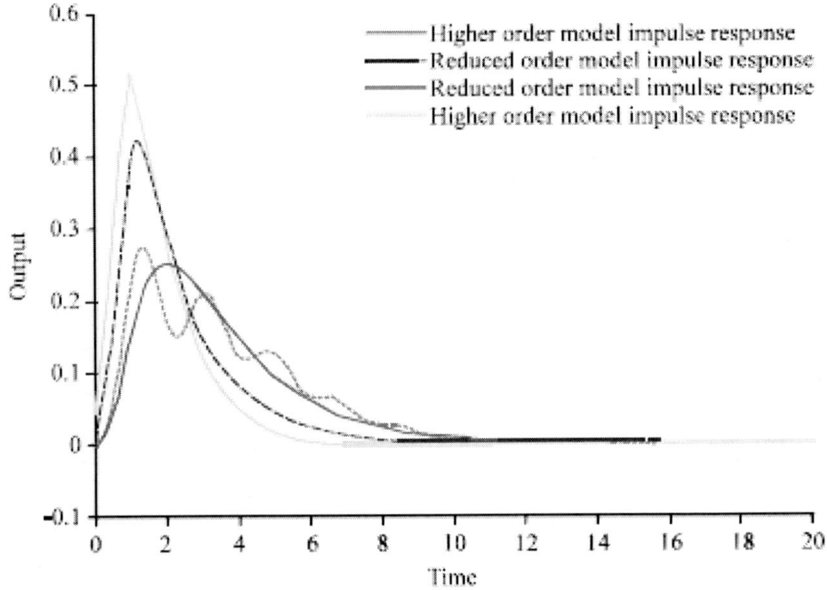

Figure 5: Comparison of impulse response of higher order and reduced order model

Table 1: Comparison of error analysis for various model order reduction

Method of order reduction	Reduced model	ISE for higher order model	ISE for lower order model
Proposed method	$\frac{[49.32, 66.6]s + [169.24, 229.64]}{[27.26, 30.13]s^2 + [92.937, 10272]s + [57.352, 63.789]}$	0.808	0.8215
Devender Kumar Saini method	$\frac{[634.7, 429.7]s + [271.7, 293.2]}{[61.5, 68.99]s^2 + [255.7, 347.1]s + [83.3, 87.67]}$	0.062	4.927
B. Bandyopadhyay	$\frac{[1.16, 1.84]s + [0.27, 0.53]}{s^2 + [0.52, 0.83]s + [0.08.3, 0.16]}$	2.259	5.954

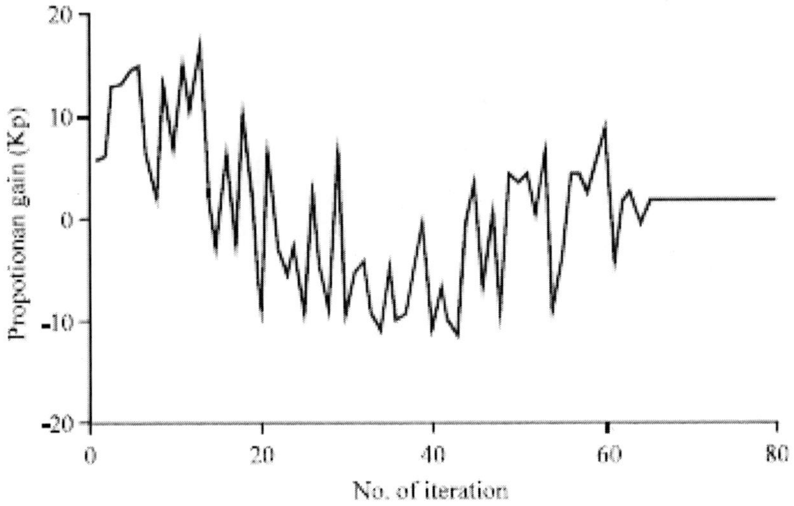

Figure 6: Convergence chart for controller setting (Kp)

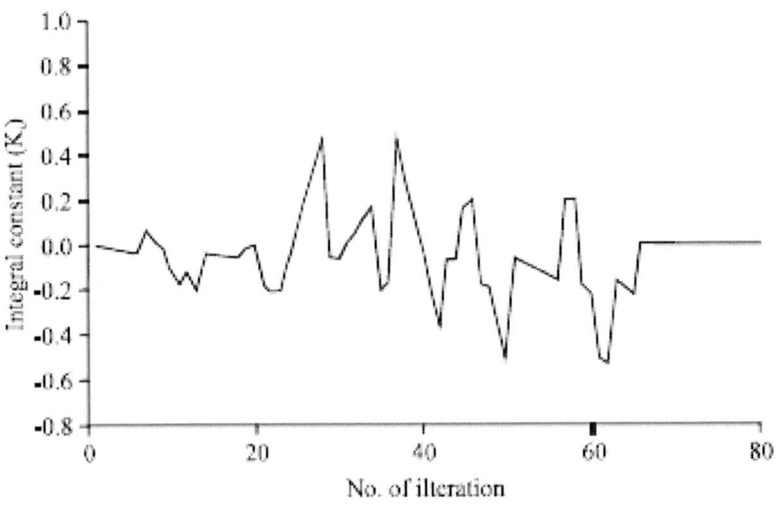

Figure 7: Convergence chart r setting (Ki)

Table 1 shows that the proposed method reduces the higher order model effectively with less error compare to other methods. Compared with Bandyopadhyay and Devender Kumar Saini method, the integral squared error of the proposed method is very low.

Controller Settings for the reduced order model are obtained using BF-PSO. Figure 6 is convergence chart for proportional controller parameter. It is observed that the proportional constant (Kp) is converged with 1.758 for the Kharitonov polynomial K11. The optimization algorithm searches the proportional constant in wide span, after 66th iteration it has converged to the constant value.

From the Fig. 7, it is observed that the integral constant (Ki) is converged to 0.009 for the Kharitonov polynomial K11. The optimization algorithm searches the proportional constant in wide span, after 66th iteration it has converged to the constant value.

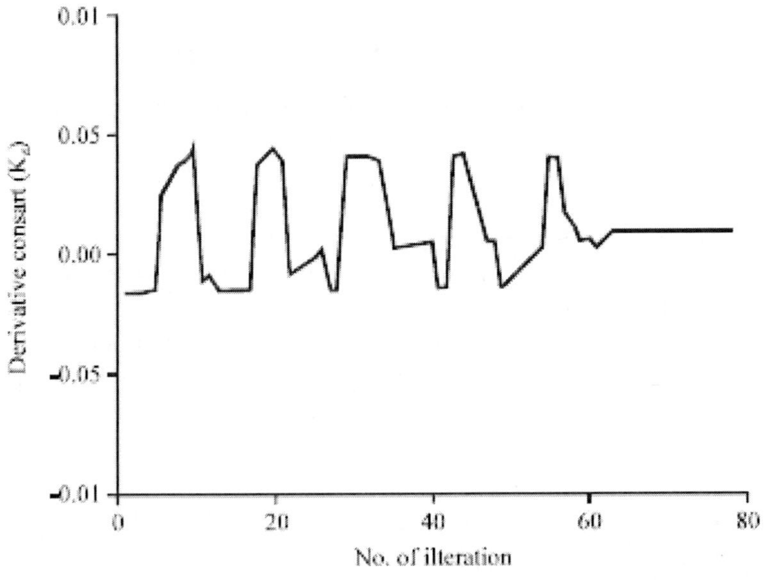

Figure 8: Convergence chart for controller setting (Kd)

From the Fig. 8, it is observed that the derivative controller Constant (Kd) is converged with Kd=0.049 for the Kharitonov polynomial K11. The optimization algorithm searches the proportional constant in wide span, after 66th iteration it has converged to the constant value.

Kp = [1.34912, 1.758] and Ki = [0.0184, 0.12528] and Kd = [0.0126, 0.5821]

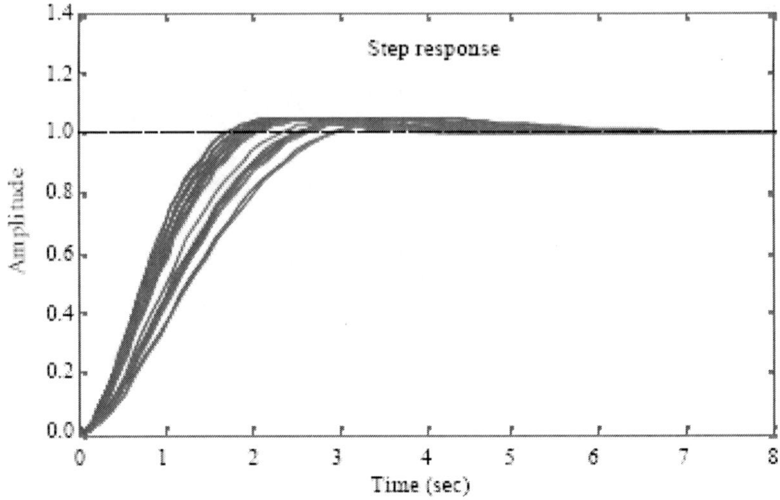

Figure 9: Step response higher and lower order model with controller

Figure 6-8 shows convergence flow of the controller setting. Figure 9 gives the comparison of the step response of higher and reduced order closed loop model. Ki = 0.0629, Kd = 0.0049 for K11 and similarly controller parameter is calculated for all other set of Kharitonov polynomials K12, K13 etc. From the set of controller parameter identified for all 16 Kharitonov polynomials, Controller parameter in interval form can be computed by using the hull set.

9. CONCLUSION

The proposed method simplifies the mathematical complexities. This model reduction method is based on convergence technique, generates stable reduced order model for original higher order stable model by retaining both initial time moments and Markov parameters. Reduced order model performance match the higher order model performance for both steady state and transient state of the time response. A controller parameter is identified for the reduced order model, using bacterial foraging based Particle swarm optimization (PSO) which control the higher order model effectively. Step and impulse responses are obtained are compared. But the time taken by the algorithm is large compared to other methods.

REFERENCES

1. Ali, A. and S. Majhi, 2006. Design of optimum PID controller by bacterial foraging strategy. Proceedings of the IEEE International Conference on Industrial Technology (ICIT), 2006, December 15-17, 2006, pp: 601-605.

2. Babu, T. and N. Pappa, 2008. Biased model reduction of discrete interval system using differentiation technique. IEEE Conf. INDICON, 2: 223-226.

3. Bandyopadhyay, B., A. Upadhye and O. Ismail, 1997. δ- γ Routh approximation for interval systems. IEEE Trans. Autom. Control, 42: 1127-1130.

4. Dolgin, Y. and E. Zeheb, 2004. Model reduction of uncertain systems retaining the uncertainty structure. Syst. Control Lett., 54: 771-779.

5. Hsu, C.C., S. C. Chang and C.Y. Yu, 2007. Tolerance design of robust controllers for uncertain interval systems based on evolutionary algorithms. IET Control Theory Appl., 1: 244-252.

6. Kennedy, J. and R. Eberhart, 1995. Particle swarm optimization. Proc. IEEE Int. Conf. Neural Networks, 4: 1942-1948.

7. Korani, W.M., H.T. Dorrah and H.M. Emara, 2009. Bacterial foraging oriented by particle swarm optimization strategy for PID tuning. Proceedings of the IEEE International Symposium on Computational Intelligence in Robotics and Automation, December 15-18, 2009, Daejeon, South Korea, pp: 445-450.

8. Lepschy, A. and U. Viaro, 1983. A note on the model reduction problem. IEEE Trans. Autom. Control, 28: 525-527.

9. Lucas, T.N., 1988. Differentiation reduction method as a multipoint pade approximation. IEE Electron. Lett., 24: 60-61.

10. Lucas, T.N., 1992. Some further observations on the differentiation method of on the model reduction problem. IEEE Trans. Autom. Control, 37: 1389-1391.

11. Passino, K.M., 2002. Biomimicry of bacterial foraging for distributed optimization and control. IEEE Control Syst., 22: 52-67.

12. Rajinikanth, V. and K. Latha, 2011. Optimization of PID controller parameters for unstable chemical systems using soft computing technique. Int. Rev. Chem. Eng., 3: 350-358.

13. Ren, B. and W. Zhong, 2011. Multi objective optimization using chaos based PSO. Inform. Technol. J., 10: 1908-1916.

14. Supriyono, H. and M.O. Tokhi, 2011. Adaptation schemes of chemotactic step size of bacterial foraging algorithm for faster convergence. J. Artif. Intell., 4: 207-219.

15. Tan, N. and D.P. Atherton, 2000. Stability and performance analysis in an uncertain world. IEE Comput. Control Eng. J., 11: 91-101.

16. Tomar, S.K., R. Prasad. S. Panda and C. Ardil, 2009. Conventional and PSO based approaches for model reduction of SISO Discrete Systems. Int. J. Electr. Electron. Eng., 2: 45-60.

17. Younseok, C., 2007. A Note on discrete interval system reduction via retention of dominant poles. Int. J. Control, Autom. Syst., 5: 208-211.

Chapter 11

Optimization of China Crude Oil Transportation Network with Genetic Ant Colony Algorithm

Yao Wang * and Jing Lu

Transportation Management College, Dalian Maritime University, Dalian 116026, Liaoning, China

ABSTRACT

Taking into consideration both shipping and pipeline transport, this paper first analysed the risk factors for different modes of crude oil import transportation. Then, based on the minimum of both transportation cost and overall risk, a multi-objective programming model was established to optimize the transportation network of crude oil import, and the genetic algorithm and ant colony algorithm were employed to solve the problem. The optimized result shows that VLCC (Very Large Crude Carrier) is superior in long distance sea transportation, whereas pipeline transport is more secure than sea transport. Finally, this paper provides related safeguard suggestions on crude oil import transportation.

KEYWORDS

integrated transportation; transportation safety; multi-objective programming; crude oil import; transportation network

1. INTRODUCTION

China has experienced an increasing demand for crude oil in recent years due to its economic boom. According to the report of International Energy Data and Analysis by the U.S. Energy Information Administration, China became the largest global energy consumer in 2011 and is the world's second-largest oil consumer behind the United States. The country was a net oil exporter until the early 1990s and became the world's second-largest net importer of crude oil and petroleum products in 2009. In 2012, China imported 271 million tons of crude oil in total, mostly from The Middle East, Africa, Russia and the Asian-Pacific. However, 90% of imported oil is carried by maritime transportation; the risk factors en route and the single mode of transportation place constraints on China's oil supply. Thus, the construction of a reliable crude oil import network is important for the safety of the national crude oil import channel.

In the past few decades, the safety of crude oil transport and optimization of the transportation network have received much attention. Douligeris *et al.* [1] described the development model of US oil maritime transportation; risk assessment and response measures for oil spills and pollution are analysed in detail. Hamzah and Basiron [2] made suggestions for maritime transportation channel safety through analysis of security factors for maritime transportation in the Malacca strait area. David [3] and other Western scholars devoted more attention to the phenomenon of piracy. They discussed the threat to sea channels from piracy and proposed a number of measures to prevent piracy and armed robbery. In addition, the International Maritime Bureau is advised to play a more important role so that the countries involved can come together to address the piracy crisis. Asian scholars are more concerned about the safety of international shipping channels within their area. Ji [4] reviewed the security status of the Straits of Malacca and the maritime transportation channel of the Asia-Pacific, respectively. Safety factors were analysed and were divided into categories; in addition, proposals and ideas on the protection of maritime transportation security were provided. Valencia [5] devoted more attention to the dispute-prone areas, such as the South China Sea, and provided relevant policy recommendations. As the political and cultural environment of the oil export countries has an impact on transportation safety, some researchers have also shown interest for the impact of diversification of crude oil imports on transportation safety. Lesbirel [6] proposed diversifying oil imports to reduce risk. Leiby [7] suggested that oil import diversification can reduce the crude oil transportation risk caused by disasters, terrorist attacks, war, regime change and so on. Related to this research content, Alhajji [8] summarized the energy security policies in developed countries: supplier diversity, source of diversification, reducing dependence on Middle Eastern oil and reducing volatility. Konoplyanik [9] added that diversification does not mean the diversification of energy suppliers. In terms of the research purpose, Vivoda [10] and other researchers demonstrated the importance of diversity to the importing country's safety using as examples the United States, China, Japan and other developed countries. Stringer [11] extended the issue; he distinguished the diversification of energy sources and the diversification of suppliers. With respect to the diversification of imports, Jia *et al.* [12] systematically classified the safety factors influencing the national crude oil import maritime transportation channel and provided advice for factors combing national policy. Wang [13] from Dalian Maritime University applied the Back Propagation Nerve Network to conduct a risk assessment on national crude oil import maritime transportation.

Concerning the optimization of crude oil transportation, Neiro and Pinto [14] established a mathematical optimization model for the operation of the oil supply chain. Iakovo [15] set up a multi-objective model of maritime oil transport route optimization and risk minimization, which achieves the optimization of both transportation routes and risk. Chen and Zhang [16] from Shanghai Maritime University adopted a non-linear programming method to optimize the network of crude oil imports dominating maritime transportation routes. Chu [17] from Dalian Maritime University built a multi-objective programming model based on the objects of minimized transportation cost and risk to analyse the network of crude oil maritime transportation in batch.

As mentioned above, the research on crude oil import transportation has thus far mainly focused on maritime transportation with no consideration for other methods, *i.e.*, pipelines. This article covers both maritime and pipeline transportation of crude oil imports. On one hand, from the perspective of quality, this article analyses the risk factors for the national crude oil import transportation channel. On the other hand, from the perspective of quantity, it sets up an integrated layout that is also the basis for the advice on safety raised herein through the multi-objective programming model with minimized total costs and risks.

This paper is organized as follows: Section 2 describes the current national crude oil import transportation channel.Section 3 analyses the risk factors for the national crude oil import channel. Section 4 proposes a multi-objective programming model for optimizing the national crude oil transportation network. Section 5 provides solution details and results analysis. Section 6 provides the conclusion.

2. CURRENT NATIONAL CRUDE OIL IMPORT TRANSPORTATION CHANNEL

A reliable oil import transportation network is of great importance to the security of China's oil imports. China has three ways of transporting crude oil. First, it is transported by sea. This is mainly for oil-producing countries and regions, which are far from China and have no land borders, such as the Middle East, Africa. Second, the oil is transported by pipeline to China. This model is mainly applied to oil exporting countries, such as Russia, Kazakhstan and other countries in Central Asia that neighbour China. Third, oil can be transported by rail, which is convenient to neighbouring countries and regions with railway connections to China, such as Mongolia and the Russian Far East. Figure 1 shows the three modes import oil transportation.

2.1. Maritime Transportation

More than 90% of China's crude oil imports is by maritime transportation. Currently, there are three primary routes: (1) Persian Gulf-Hormuz Strait-Malacca Strait-China; (2) North Africa-Mediterranean-Strait of Gibraltar-Cape of Good Hope-Malacca Strait-China; and (3) West Africa-Cape of Good Hope-Malacca Strait-China, as shown in Table 1.

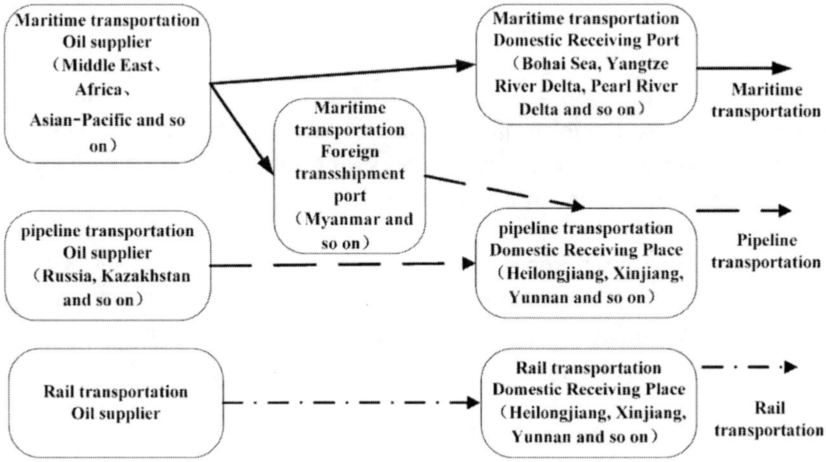

Figure 1. Forms of Transportation for Chinese Oil Imports.

The Malacca Strait, the Taiwan Strait, the Bab el-Mandab, the Strait of Hormuz, the Suez Canal, *etc.*, have a significant impact on China's oil import routes. The Strait of Hormuz is the world's most important oil chokepoint because of its daily oil flow of 17 million barrels per day in 2013. Energy Information Administration estimates that more than 85% of the crude oil that moved through this chokepoint went to Asian markets, most of which went to China's. Thus, with the rapid growth in China's oil imports, the strategic position of the Strait of Malacca in China's oil security has become increasingly prominent. It can be seen that importing by sea is relatively simple, with a high degree of dependence on the Malacca Strait and Cape of Good Hope.

Table 1. Routes of China's crude oil imports by maritime transportation.

Routes	Details of the Routes
Persian Gulf	Persian Gulf-Hormuz Strait-Malacca Strait-China
North Africa	North Africa-Mediterranean-Strait of Gibraltar-Cape of Good Hope-Malacca Strait-China
West Africa	West Africa-Cape of Good Hope-Malacca Strait-China

2.2. Pipeline

Pipeline transportation is a major mode of importing oil, mainly for transport over land, which has the advantages of large transport volumes, safety, convenience, economy, stability and others. It is an important complement to transportation by sea.

With the completion of the China-Myanmar pipeline project, three strategic crude import inland channels have been formed and situated in the Northwest, Northeast and Southwest. In the Northwest channel, there is the China-Kazakhstan pipeline project and the Middle Asia natural gas pipeline project; in the Northwest channel, the China-Russia crude and gas pipeline; and in the Southwest channel, the China-Myanmar crude and gas pipeline.

2.3. Rail

Rail transport is an important supplement of marine transportation and pipeline transportation. The volume of rail transportation is limited, and transportation costs are higher, but, sometimes, maritime transportation is not viable or there is no pipeline, and rail transportation is thus the only option. In the long term, rail transportation is not expedient.

At present, there are two lines for rail imports; one is Russian Irkutsk-China Manzhouli, and the other is Kazakhstan-China Xinjiang Alashankou. Since the opening of the pipeline, the rail volumes have gradually fallen.

3. ANALYSIS OF RISK FACTORS FOR THE NATIONAL CRUDE OIL IMPORT CHANNEL

3.1. Risk Factors for the Maritime Transportation Channel

To provide an overview of the previous studies [18,19,20,21,22,23], the safety of maritime transportation is affected by supply risk, price risk, transportation risk and other risk. This paper gives more attention to the supply risk, transportation risk. Thus, the first tier of risk factors can be summarized. The risk factors of war and regional conflict, geopolitics and international situation, import source stability represent the supply risk, and the other risk factors of terrorism and pirates, transportation distance, weather and sea state, traffic capacity refer to the transportation risk. In order to explain the first tier of risk factors preferably, this paper gives seventeen second tier of risk factors. The risk factors of the maritime transport channel are shown in Table 2.

Table 2. Risk factors for marine transportation of crude oil imports.

Risk Factors (First Tier)	Risk Factors (Second Tier)
war and regional conflict(A)	racism (A1) resource competition (A2) sea territory conflict (A3)
geopolitics and international situation (B)	geopolitics change (B1) international conflict change (B2)
terrorism and pirates (C)	seaborne terrorist attack (C1) terrorist active area en route (C2)
import source stability (D)	monopoly of crude supply (D1) international cooperation (D2) maritime conventions (D3)
transportation distance (E)	passage (E1)
weather and sea state (F)	severe weather condition (F1) sea geography (F2) sea chart and document preparation (F3)
traffic capacity (G)	strait or canal en route (G1) island en route (G2) ship density en route (G3)

3.2. Risk Factors of Pipeline Transportation Channel

Factors affecting the safety of oil pipeline transportation are derived from the following aspects: (1) safety of the pipeline, mainly the design, construction procedures and defects after application; stress-corrosion cracking (SCC) for the underground pipe is one of the most important factors of the pipeline system. (2) Sabotage is usually a safety concern of third parties, *i.e.*, theft of oil, damage to construction projects on the ground. (3) Natural disasters and force majeure, a non-ignorable factor.

Hurricane Katrina caused heavy damage and interrupted crude and gas production in the US Gulf.

4. OPTIMIZING THE NATIONAL CRUDE IMPORT TRANSPORTATION NETWORK UNDER THE MULTI-OBJECTIVE PROGRAMMING MODEL

4.1. Assumptions

(1) The research period is set to 1 year.

(2) Take no consideration of crude purchase costs and only focus on optimizing transportation costs.

(3) Risk for the crude import transportation channel in this article only refers to pirate attacks in the Malacca Strait.

(4) Include the China-Myanmar pipeline project in the model because of its importance.

4.2. Multi-Objective Programming Model

The goal of this paper is to minimize the total travel cost and overall risk; thus, there are two objectives. This paper therefore applies the multi-objective programming model.

Objective function 1: Minimize the total annual cost of the crude transportation network. Annual total cost includes the transportation costs from the supply port via seaborne transportation to port transhipment or receiving port, costs from port transhipment or receiving port via pipeline to the final destination of consumption, and costs from the supply side via pipeline to consumption.

$$Min\ \ Z_1=\sum_{i=1}^{p}\sum_{j=1}^{q}\sum_{k=1}^{l}C_{ij}^{k}\times X_{ij}^{k}+\sum_{i=p+1}^{m}\sum_{r=1}^{u}C_{ir}'\times X_{ir}'+\left(\sum_{i=1}^{p}\sum_{t=1}^{v}\sum_{k=1}^{l}C_{it}^{k}\times X_{it}^{k}+\sum_{t=1}^{v}\sum_{r=u+1}^{n}C_{tr}'\times X_{tr}'\right)$$
$$=\sum_{i=1}^{p}\sum_{j=1}^{q+v}\sum_{k=1}^{l}C_{ij}^{k}\times X_{ij}^{k}+\sum_{i=p+1}^{m+v}\sum_{r=1}^{n}C_{ir}'\times X_{ir}' \tag{1}$$

Objective function 2: Minimize the risk from crude oil supply risk and seaborne transportation risk.

$$Min\ \ Z_2=\sum_{i=1}^{p}\xi_i'\sum_{j=1}^{q+v}\sum_{k=1}^{l}X_{ij}^{k}+\sum_{i=1}^{p}\sum_{j=1}^{q+v}\sum_{k=1}^{l}\xi_{ij}^{k}X_{ij}^{k} \tag{2}$$

$$\sum_{j=1}^{q+v}\sum_{k=1}^{l}X_{ij}^{k}\le Q_i,\ i=1,2,\ldots,p \tag{3}$$

$$\sum_{r=1}^{u}X_{ir}'\le Q_i,\ i=p+1,\ldots,m \tag{4}$$

$$\sum_{i=1}^{p}\sum_{k=1}^{l}X_{ij}^{k}\le F_j,\ j=1,2,\ldots,q,\ldots,q+v \tag{5}$$

$$\sum_{i=1}^{p}\sum_{t=1}^{v}\sum_{k=1}^{l}X_{it}^{k}=\sum_{t=1}^{v}\sum_{r=u+1}^{n}X_{tr}' \tag{6}$$

$$X_{ij}^{k}\ge 0,\ i=1,2,\ldots p,\ j=1,2,\ldots,q+v,\ k=1,2,\ldots,l \tag{7}$$

$$X_{ir}'\ge 0,\ i=p+1,\ldots,m+v,\ r=1,2,\ldots,n \tag{8}$$

where I is the aggregation of the crude supply area, I={i∈I|i=1,2,...,p,...,m}. J is the aggregation of crude import receiving ports, J={j∈J|j=1,2,...,q}. R is the aggregation of crude import pipeline receiving places, R={r∈R|r=1,2,...,u,...,n}. T is the aggregation of crude import transhipment ports, T={t∈T|t=1,2,...,v}. K is the aggregation of crude import tanker size,

K={k∈K|k=1,2,...,l}. Xkij is the import crude volume carried by size k tanker from supply area i via seaborne transportation to receiving ports j. X′ir is the import crude volume from supply area i via pipeline to pipeline receiving places r. Ckij is the transportation unit cost of tanker size *k* carrying crude from supply area i via seaborne transportation to crude import receiving ports j. C′ir is the transportation unit cost from crude supply area i via pipeline to pipeline receiving place r. ξ′i is the risk factor of i crude supply ξkij is the transportation risk factor using size *k* tanker to carry crude imports from supply area i to crude import receiving ports j.

4.3. Calculation of Risk Factor

4.3.1. Risk Factor of Crude Oil Supply

The article takes the crude oil supply interruption as random and adopts exponential distribution f(x)=1−eax; the random variable is sustained event t and event impact r.

Table 3 shows the interruption cases of world oil supply including period, volume, percentage of production, region and last time between 1951 and 1980. Middle East is the region which has the most interruption cases. It has twelve cases between 1951 and 1980, and the other cases happened in Africa between 1970 and 1971. As for the last time of net month of supply cut off, the first case and the last case have the longest last time, they lasted for 12 and 44 month, respectively, which have a serious effect on oil supply safety, especially for the last case, because the volume of oil is large. Adopting statistical methods and through regression analysis, cumulative distribution probability density f(t) and f(r) for random variables tand r are obtained. In accordance with traits of exponential distribution, the cost factor of supply risk is

$$E\left(\xi'\right)=\frac{n}{N}\times\varphi_t\times\varphi_r\times\frac{365}{12}\times\rho \quad (9)$$

4.3.2. Risk Factor of Seaborne Transportation

This article assumes pirate attacks in the Malacca Strait as the crude import seaborne transportation risk, and transhipment via Myanmar is excluded. Further, the higher the ship freeboard is, the more difficult it is to abduct. Therefore, ship size is the best index of pirate risk assessment. *x* is the ship freeboard, and the cumulative probability function of the pirate abduction is [24]:

$$P(x) = \frac{1}{1+e^{-(ax+b)}} \tag{10}$$

In accordance with ship size and number of pirate encounters in January 2003 and December 2007, Table 4 is obtained by formula calculation.

a and b are obtained through logit alternation linear regression; and the cumulative probability density of a tanker under pirate attack p(x) is calculated in accordance with the traffic capacity of crude oil import seaborne transportation in the Malacca Strait ε the cost factor of the pirate risk for different size of tanker is:

$$\xi_{ij}^{k} = C_{ij}^{k} \times \varepsilon_{k} \times \int_{x_1}^{x_2} p(x) \tag{11}$$

5. MODEL SOLVING METHOD AND RESULT ANALYSIS

5.1. Data Preparation

Crude oil import volume by supply region (10,000 tons): (1) Middle East—11,950; (2) Africa—6690; (3) Asia Pacific—1258; (4) Russia—1849; (5) Kazakhstan—1121. Handling capacity of receiving port and transhipment port (10,000 tons): (1) Bohai Bay area—15,780; (2) Yangtze River delta area—12,300; (3) Pearl River delta area—4800; (4) Myanmar (transhipment)—2200. Crude oil import pipeline annual receiving capacity (10,000 tons) and unit cost ($/ton): China-Russia pipeline annual capacity 3000, unit cost 17.43. China-Kazakhstan pipeline annual capacity 2000, unit cost 31.78. China-Myanmar pipeline annual capacity 2200, unit cost 21.61. Crude oil import seaborne transport unit cost as shown in Table 5.

Table 3. Interruption cases of world crude oil supply.

Period	Volume/mbpd	Percentage of Production/%	Last Time (Net Month of Supply Cut Off)	Region
1951.03–1954.10	0.7	3.5	44	Middle East
1956.11–1957.03	2	9.5	4	Middle East
1966.12–1967.03	0.7	2	3	Middle East
1967.06–1967.08	2	5.4	2	Middle East
1970.05–1971.01	1.3	2.6	9	Africa
1971.04–1971.08	0.6	1.2	5	Africa
1973.03–1973.05	0.5	0.9	2	Middle East
1973.10–1974.03	2.6	4.4	6	Middle East
1976.04–1976.05	0.3	0.5	2	Middle East
1977.05–1977.06	0.7	1.1	1	Middle East
1978.11–1979.04	3.5	5.4	6	Middle East
1980.10–1980.12	3.3	5.2	3	Middle East
1990.08–1990.10	4.6	7	3	Middle East
1994.04–2000.03	3.3	7.7	12	Middle East

Table 4. Distribution list for freeboard height and cumulative probability of piracy.

Ship Size	Freeboard Range	Average	Frequency	Cumulative Frequency	Cumulative Probability
	1.5–2.0	1.75	49	49	0.1701
	2.0–2.5	2.25	22	71	0.2465
Handysize	2.5–3.0	2.75	26	97	0.3368
	3.0–3.5	3.25	28	125	0.4340
	3.5–4.0	3.75	53	178	0.6181
Panamax	4.0–4.5	4.25	31	209	0.7257
	4.5–5.0	4.75	16	225	0.7813
Aframax	5.0–5.5	5.25	21	246	0.8542
	5.5–6.0	5.75	21	267	0.9271
	6.0–6.5	6.25	2	269	0.9340
Suezmax	6.5–7.0	6.75	9	278	0.9653
	7.0–7.5	7.25	0	278	0.9653
	7.5–8.0	7.75	0	278	0.9653
VLCC	8.0–8.5	8.25	1	279	0.9688
	8.5–9.0	8.75	7	286	0.9931
ULCC	9.0–9.5	9.25	2	288	1.0000

Table 5. Unit cost of shipping transport (unit: US dollar per ton).

Size \ Route	1-1	1-2	1-3	1-4	2-1	2-2	2-3	2-4	3-1	3-2	3-3	3-4
Panamax	/	/	/	/	/	/	/	/	12.64	10.78	8.89	/
Aframax	23.47	21.81	19.59	13.2	/	/	/	/	9.85	8.41	6.93	/
Suezmax	17.07	15.86	14.24	9.6	/	/	/	/	/	/	/	/
VLCC	11.24	10.45	9.38	6.32	17.3	16.5	15.4	12.57	/	/	/	/

5.2. Model Solving—The Genetic Ant Colony Algorithm

The basic idea of the model solution is to transform the double goal programming model into a single objective programming model by a linear combination treatment. Double target network optimization problem solving has been a more difficult problem class; according to the research results, this type of problem is NP (Nondeterministic Polynomial)-Hard, so the type of classical optimization model is powerless. This paper constructs a genetic algorithm based on the following genetic colony ant algorithm for the reference of the relevant research results, as shown in Table 6.

The above algorithm is basically the same as the traditional genetic algorithm process, the only difference is the following: When calculating the fitness, we use the linear programming method to optimally calculate the different modes of transportation volume. This approach allows us to not have to think much about how to express the design algorithm coding but only to consider the encoding method for the route network. Below, we highlight the encoding method that is the most central part of the algorithm: the route network.

Unlike other water transport networks, oil transportation networks are usually relatively simple; two end points with a roundtrip transport is its main mode. For this feature, we design the coding method as shown in Figure 2. Each individual coding is constituted of two parameters, where parameter 1 is used to represent the oil exporter choice. For example, in this case, oil exporter 1 and 2 are selected, so that the gene loci bits 1 and 2 of parameter 1 are 1; because oil exporter 3 is unchecked, the gene bit 3 of parameter 1 is 0. The selection mechanism of Parameter 2 is similar to Parameter 1, and its function is to filter the oil importer. For example, in Figure 2 the ports 1, 2, 3 are selected so that the corresponding funds bit is 1, while port 4 is unchecked, so the gene loci bits 4 is 0 of parameter 2.

The final question is how to link the exporter and importer. Actually, this problem is a VRP (Vehicle Routing Problem) problem; the only difference between our problem and the classic one is that we need to consider the oil supply capacity of each exporter. In view of this, we reference Zhan [25] and use the ant colony algorithm to solve this problem.

Table 6. The Genetic Process of Model Solving.

Step	Content
Step 1	Randomly generate the initial population of the shipping line programs
Step 2	For each individual program k, the original optimization model is equivalent to a linear programming model; to solve the problem under program k conditions, obtain the lowest oil import program to get (existing traffic) $X_{ij}^{k\beta}$, $X'_{ir\beta}$ and the minimum cost of the program. This (generalized cost transportation system) $z = \alpha z_1 + \beta z_2$ is the individual fitness of program k
Step 3	Perform a crossover and mutation operation to obtain the new population
Step 4	Check whether the genetic algorithm termination condition is satisfied: $\lvert \min(z_k) - avg(z_k) \rvert < \varepsilon$, where $\min(z_k)$ represents the maximum value of the current population of individual fitness, $avg(z_k)$ represents the average individual fitness of the current population, and ε is the termination threshold. If the termination condition is satisfied, the algorithm terminates, else perform Step 2.

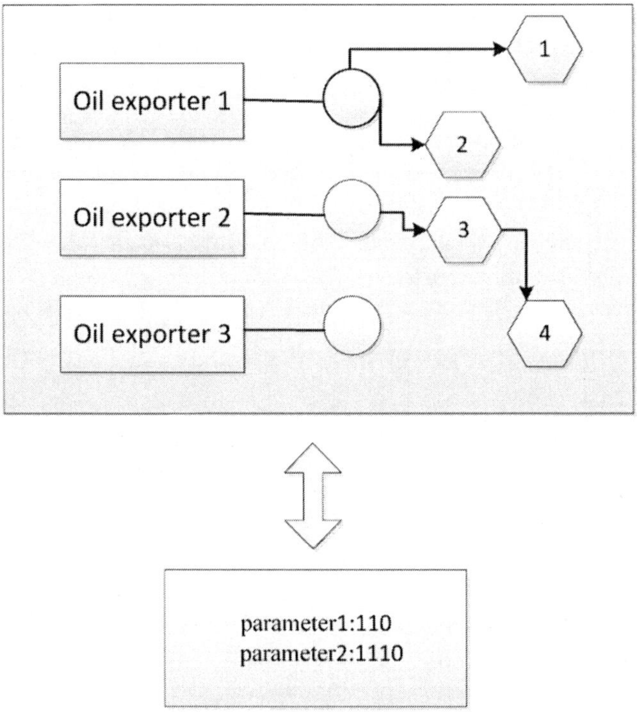

Figure 2. Diagram of Coding Method.

The ant colony algorithm is inspired by the true nature of ant colony foraging behaviour, which was first proposed in 1992 by Italian scholars Dorigo M. It is a heuristic algorithm that can be used to solve combinatorial optimization problems. After many studies, researchers found that individual ants transfer information through a substance called pheromone. In the course of movement, ants can leave pheromones in their path and tend to move in the direction of a high intensity of pheromones. Based on the principle of the ant colony algorithm and Zhan's ideas [25], the specific algorithm is as follows:

When the algorithm starts, the ants randomly distribute in the oil export paces; then, according to Equation (12), the savings between every two oil exporters is calculated, which expresses the cost savings by merging two oil exporters; then, according to Equation (13), the transition probability is calculated, which refers to the likelihood of the ant to select oil export h after the selection of g. Finally, let the ants search a path in two phases according to the probability: The first phase is from the start of oil exports to the search for oil imports, and the second stage is from the oil imports to the search for and selection of oil exporters.

$$q_{gh} = d_{gm} + d_{hm} - d_{gh} \tag{12}$$

q represents savings, d represents the cost, m is on behalf of oil imports, and g,h represents oil exporters.

$$p_{gh} = \begin{cases} \frac{[\tau_{gh}]^{\partial}[q_{gh}]^{\beta}}{\sum_{s\in\Psi}[\tau_{gh}]^{\partial}[q_{gh}]^{\beta}}, l_g \in \Psi, \Psi = \{l_g \in L : l_g\} \\ 0, \text{otherwise} \end{cases} \tag{13}$$

τgh represents pheromones; α, β are two parameters used to represent the importance of pheromones and savings.

After all ants complete the path construction, according to Equation (14), the pheromone is updated.

$$\tau'_{gh} = (1-\rho)\tau^{s}_{gh} + \sum_{\mu=1}^{\lambda-1} \Delta\tau^{\mu}_{gh} + \lambda\Delta\tau^{*}_{gh} \tag{14}$$

1 ≤ ρ ≤ 0 represents the evaporation rate of the pheromone; λ represents the number of elite ants.

However, it should be noted that here we allow the existence of a link with the oil import port, in this case port 4; as the oil output place through port 3 to 4 is a close path, we will connect port 3 and port 4.

By the above genetic algorithm and utilizing MATLAB software, a fuzzy optimal solution is obtained for the crude oil import network: total cost of crude oil import transportation 332,060 ($10,000) and total risk 45,693. The transportation plan is shown in Table 7.

In consideration of ship load and transportation feasibility, adjustment to the aforesaid plan is achieved by setting a minimum loadable for each size; the result is the total cost of crude oil import transportation 339,620 ($10,000), total risk 45,576. The adjusted transportation plan is as shown in Table 8.

Table 7. Transport plan of China crude oil import.

		Bohai Area	Yangtze River Delta Region	Pearl River Delta Region	Myanmar (Transhipment)	Total (10,000 tons)
Middle East	Aframax	0.2	0.2	0.2	1	1.6
	Suezmax	0.4	0.4	0.7	1.1	2.6
	VLCC	4866.5	5267.9	1805.9	5.5	11,945.8
Africa	VLCC	1949.4	2208.8	1753.4	778.4	6690
Asia	Panamax	0.4	0.5	0.7	/	1.6
Pacific	Aframax	2.9	19.7	1233.8	/	1256.4
Total (10,000 tons)		6819.8	7497.5	4794.7	786	19,898
China-Russia		/	/	/	/	1849
China-Kazakhstan		/	/	/	/	1121
China-Myanmar		/	/	/	/	786

Table 8. Adjusted transport plan of China crude oil imports.

		Bohai Area	Yangtze River Delta Region	Pearl River Delta Region	Myanmar (Transhipment)	Total (10,000 tons)
Middle East	Aframax	0	0	0	0	0
	Suezmax	0	0	0	0	0
	VLCC	4690.9	5428.5	1830.6	0	11,950
Africa	VLCC	1941.7	1923.3	1711.4	1113.6	6690
Asia	Panamax	0	0	0	/	0
Pacific	Aframax	0	0	1258	/	1258
Total(10,000 tons)		6632.6	7351.8	4800	1113.6	19,898
China-Russia		/	/	/	/	1849
China-Kazakhstan		/	/	/	/	1121
China-Myanmar		/	/	/	/	1113.6

5.3. Result Analysis

Based on Table 7 and Table 8, the following results can be determined:

(1) In terms of transportation costs and ship load considerations, VLCC is widespread in the Middle East and Africa routes, which illustrates that the lower cost of VLCC is superior in long distance seaborne transportation. This type of tanker dominates the deployments.

(2) The Aframax and Suezmax tankers from the Eastern-Bohai and the Middle East-Airline Yangtze River Delta region only carry 0.2 and 0.4 million tons, respectively; the Middle East-Pearl triangle routes only carry 0.2 and 0.7 million tons, and the Middle East-Myanmar port transit only carries 1 and 1.1 million tons, which is of little practical significance; these shipments can be ignored and can be supplemented in other places. Similarly, the traffic of VLCC in the Middle East-Myanmar transit, Panamax and Aframax in the Asia-Pacific routes are also very little, which can be ignored.

(3) The Bohai area and Yangtze River Delta region mainly rely on Middle East and African crude oil imports, and discharge volume accounts for approximately 70% of total crude import seaborne transportation at the national import crude receiving centre.

(4) Transit traffic in the port of Myanmar is from African routes, not Middle East routes, because oil imports from Africa are smaller than those from the Middle East. This shows that in the case of large traffic, the cost for pipeline transportation is also much larger than that of maritime transport.

(5) The adjusted crude import transportation plan sharply increases the import volume via the China-Myanmar pipeline with a lower total risk, despite higher total transportation costs, which demonstrates the safety of pipeline transportation.

6. CONCLUSIONS

Based on the depth analysis of risk factors for China's crude oil import transportation channel and rationalization assumptions, a multi-objective programming model for China's crude oil import transportation network is established; its goal is to minimize the total transport cost and overall risk. Based on the genetic algorithm and ant colony algorithm as well as MATLAB software, an optimized network transportation program for crude oil imports is provided. The optimization results show that, to protect the safety of China's crude oil import channel, the following conditions must be met:

On one hand, a rational layout for a large scale crude terminal meeting berth and discharge of VLCC-size and larger tankers will in the future improve the receiving process for crude imports. On the other hand, the construction of a diversified crude oil supply source and import channel can reduce the impact of regional political risks and regional conflicts on national crude imports, which will ensure supply stability. In addition, a new transportation channel for crude imports should be explored; land transportation should be enhanced to ease the safety concerns that stem from seaborne transportation; and a multi-layer crude import transportation network should also be established.

ACKNOWLEDGMENTS

This work was supported by National Natural Science Foundation of China 71303026, 71302044 & 71473023, the Projects of Philosophy and Social Sciences from Liaoning Province (L13CJY049).

AUTHOR CONTRIBUTIONS

Both authors contributed extensively to the work presented in this paper. Both authors have read and approved the final manuscript.

CONFLICTS OF INTEREST

The authors declare no conflict of interest.

REFERENCES

1. Douligeris, C.; Iakovou, E.; Englehardt, J.D.; Li, H.; Ip, C.M.; Mooers, C.N.K. Development of a national marine oil transportation system model. *Spill Sci. Technol. Bull.* **1997**, *4*, 113–121.

2. Hamzah, B.A.; Basiron, M.N. The Straits of Malacca: Some Funding Proposals. In *MIMA Paper*; Maritime Institute: Kuala Lumpur, Malaysia, 1997; p. 67.

3. Piracy and Armed Robbery Against Ships. Available online: http://www.imo.org/OurWork/Security/PiracyArmedRobbery/Pages/Default.aspx (accessed on 31 March 2001).

4. Ji, G. *SLOC Security in the Asia-Pacific*; Asia-Pacific Center for Security Studies: Honolulu, HI, USA, 1999.

5. Valencia, M.J. Sea of Japan: Transnational marine resource issues and possible cooperative responses. *Mar. Policy* **1990**,*14*, 507–525.

6. Lesbirel, S.H. Diversification and energy security risks: The Japanese case. *Jpn J. Polit. Sci.* **2004**, *5*, 1–22.

7. Leiby, P.N. *Estimating the Energy Security Benefits of Reduced US Oil Imports*; Oak Ridge National Laboratory: Oak Ridge, TN, USA, 2007.

8. Alhajji, A.F. What is energy security? Definitions and concepts. *Middle East Econ. Surv.* **2007**, *50*, 5.

9. Konoplyanik, A. The view from Brussels. In *Emerging Threats to Energy Security and Stability*; Springer: Berlin, Germany, 2005; pp. 79–86.

10. Vivoda, V. Diversification of oil import sources and energy security: A key strategy or an elusive objective? *Energy Policy* **2009**, *37*, 4615–4623.

11. Stringer, K.D. Energy security: Applying a portfolio approach. *Balt. Secur. Def. Rev.* **2008**, *10*, 121–142.

12. Jia, D.S.; Sun, J.Y.; Luo, H.B. Chinese importing oil marine lane safety evaluation and strategy. *J. Dalian Marit. Univ.***2006**, *2*, 62–66.

13. Wang, B. Analysis on the Safety of SLOCS for Crude Oil Import of China. Master's Thesis, Dalian Maritime University, Dalian, China, June 2009.

14. Neiro, S.M.; Pinto, J.M. A general modeling framework for the operational planning of petroleum supply chains.*Comput. Chem. Eng.* **2004**, *28*, 871–896.

15. Iakovou, E.T. An interactive multiobjective model for the strategic maritime transportation of petroleum products: Risk analysis and routing. *Saf. Sci.* **2001**, *39*, 19–29.

16. Chen, F.E.; Zhang, R.Y. Optimization of Chinese crude oil import route network. *J. Shanghai Marit. Univ.* **2006**, *27*, 75–80.

17. Chu, L.Y. Study on Waterborne Petroleum Logistics System and Distribution & Transportation Network of China. Ph.D. Thesis, Dalian Maritime University, Dalian, China, March 2007.

18. Neiro, S.M.S.; Pinto, J.M. A general modeling framework for the operational planning of petroleum supply chains.*Comput. Chem. Eng.* **2004**, *28*, 871–896.

19. Dempster, M.A.H.; Pedrón, N.H.; Medova, E.A.; Scott, J.E.; Sembos, A. Planning logistics operations in the oil industry. *J. Oper. Res. Soc.* **2000**, *51*, 1271–1288.

20. Al-Otheman, W.B.E.; Lababidi, H.M.S.; Alatiqi, I.M.; Al-Shayji, K. Supply chain optimization of petroleum organization under uncertainty in market demands and price. *Eur. J. Oper. Res.* **2007**, *89*, 822–840.

21. Alcantara, V.; Duarte, R. Comparison of energy intensities in European Union countries, results of a structural decomposition analysis. *Energy Policy* **2004**, *2*, 177–189.

22. Konoplyanik, A. The View from Brussels. In *Emerging Threats to Energy Security and Stability*; McPherson, H., Wood, W.D., Robinson, D.M., Eds.; Springer: New York, NY, USA, 2005; pp. 79–86.

23. Stringer, K.D. Energy security: Applying a portfolio approach. *Balt. Secu. Def. Rev.* **2008**, *10*, 121–142.

24. Shi, J.D. The Risk Assessment on the Piracy in Malacca-Singapore Straight. Master's Thesis, Dalian Maritime University, Dalian, China, June 2009.

25. Zhan, Y.H.; Wu, Q.J. Research on Ant Colony Optimization (ACO) for the Vehicle Routing Problem. *Comput. Eng. Sci.***2008**, *30*, 60–62.

Chapter 12

Identifying Travel Mode with GPS Data Using Support Vector Machines and Genetic Algorithm

Fang Zong *, Yu Bai, Xiao Wang, Yixin Yuan and Yanan He

College of Transportation, Jilin University, Changchun 130022, China

ABSTRACT

Travel mode identification is one of the essential steps in travel information detection with Global Positioning System (GPS) survey data. This paper presents a Support Vector Classification (SVC) model for travel mode identification with GPS data. Genetic algorithm (GA) is employed for optimizing the parameters in the model. The travel modes of walking, bicycle, subway, bus, and car are recognized in this model. The results indicate that the developed model shows a high level of accuracy for mode identification. The estimation results also present GA's contribution to the optimization of the model. The findings can be used to identify travel mode based on GPS survey data, which will significantly enhance the efficiency and accuracy of travel survey and data processing. By providing crucial trip information, the results also contribute to the modeling and analyzing of travel behavior and are readily applicable to a wide range of transportation practices.

KEYWORDS

Global Positioning System (GPS); travel survey; travel mode; Support Vector Classification; genetic algorithm

1. INTRODUCTION

Travel surveys are one of the most important ways of obtaining critical information needed for transportation planning and decision making. Traditionally, a travel survey used to be conducted using different methods such as telephone/face-to-face interviews and computer-based reporting to maintain a diary [1]. These have proven to be a burden for participants to use, as well as being expensive and time consuming [2]. In addition, in these surveys, respondents often miss short trips, and round up the travel mode and travel time. The GPS-based travel survey can address this problem by recording travelers' GPS traces automatically. With the advantages of reducing respondents' participant burden and increasing data accuracy, GPS-based travel surveys have been conducted in many large cities, such as Beijing and New York. Nevertheless, GPS records cannot provide us with trip information that can be applied directly when analyzing and modeling travel behavior. In order to obtain the needed trip information, several major modeling steps, e.g., trip identification, mode detection, and travel time determination, have to be conducted based on the raw GPS data. This paper will present a model for travel mode detection, which is one of the crucial steps in trip information identification. Compared with the previous studies, it will try to enhance the identification accuracy by employing the SVC as well as using more GPS records.

The remainder of this paper is organized as follows. In Section 2, a review of identification of travel mode with GPS data in general is presented. Section 3 is a description of the data available for the study. This section is followed by a presentation of SVC and GA in Section 4. Section 5 presents the constructing process and modeling results of the mode detection model. The paper closes with some major conclusions and a discussion of future research directions.

2. EXISTING LITERATURES

Being a crucial step in GPS-based travel behaviors identifying, mode recognition has been investigated by many studies. Some researchers designed a threshold for parameters to detect travel mode, which is also called the criteria-based method. For example, Gong *et al.* [3] detected walking, subway, bus, and car by setting criteria for variables of speed and acceleration,*etc.*, with the GPS data collected in New York. Liu and Zheng [4] used parameters including average speed and median speed to identify walking, bicycle, car, bus, and subway. Besides GPS data, some studies also introduced Geographic Information System (GIS) technology to recognize travel modes, especially bus and subway. For example, GIS and actual road network information were utilized in Stopher and Greaves's study [1] for bus and subway recognition. Bohte and Maat [5] introduced a method combining GPS data, GIS information, and a web-based validation application to identify walking, bicycle, car, and railway.

Most of the previous studies conducted mode identification by building detection models. A lot of methods in this category, including neural network [6,7], decision tree [8,9,10], Bayesian network [9,11,12], Support Vector Machines (SVM) [9,13,14], and conditional random field [9], have been applied in detecting travel mode. Most of the input variables come from the GPS data itself. In these studies, Zhang *et al.* [13] presented an SVM model, which achieved a relatively higher average accuracy rate, which proves the better performance of SVM than other methodologies in mode detection.

Although the previous studies proposed many methods for mode identification, most of these studies' sample sizes were quite small. For example, Zhang [15] used the GPS data collected from 23 volunteers, Gong *et al.* [3] employed 35 respondents' GPS traces, Vij *et al.* [16] conducted the mode detection by using 11 travelers' GPS data, and Du and Aultman-Hall [17] monitored 12 volunteers' GPS traces. The inadequate data sample size may limit the procedure of model construction and estimation, and turn out to impact the performance of mode identification. Moreover, mode identification in big cities is more complicated due to complex traffic conditions and infrastructures, like urban canyons, bus routes, and subway networks. A large sample size can benefit data processing so as to provide more records for data analysis and model constructing. Bolbol *et al.* [18] took advantage of the GPS data of a large-scale study conducted in the Netherlands in 2007, which collected 1104 respondents' one-week survey data. Their results indicated that, for mode detection in a one-day GPS-based travel survey, the sample size needed is 289 and 271, respectively, for bus and car trips. By enhancing the sample size of GPS data, the accuracy rate of mode detection can be further enhanced. Therefore, this paper will construct the mode detection by using GPS survey data collected from 900 respondents, which is also part of a large-scale OD survey in Beijing, 2010. In order to enhance the performance of mode identification, SVC will be employed to establish the detection model.

3. DATA

In this section the major information on survey methods and data processing is introduced. Beijing is chosen as the study area of this paper. Being the capital city of China, it has an integrated urban land use and composite transportation network, which includes a complex road network and one of the largest public transit systems in the world. The composite land use and transportation conditions make the data survey and mode detection very complicated. The problem of signal loss due to urban canyon and subway trips also has to be addressed in data processing.

3.1. Data Survey

This study takes advantage of GPS survey data collected as part of a large-scale OD survey conducted in Beijing in 2010. The data sample consists of 900 respondents' more than one million travel records collected during the survey period from 20 October 2010 to 26 November 2010. Each respondent has an average of about 1100 records. The survey area was 16,410.54 km2 and the residing population of Beijing in 2010 was about 19.6 million. During the survey period, in addition to one-day GPS records, the 900 respondents also reported their travel diaries and socio-demographic information by filling in a paper form. With records containing missing values eliminated, our final sample consists of 1,872,431 GPS records.

Each record in the dataset represents a GPS signal that was captured consecutively at the 5-s interval by the GPS device (i-gotU GT-600) and contains information on index, date and time (universal coordinated time, UTC), latitude, longitude, altitude (m), speed (m/h), course (°), distance (m), EHPE (estimated horizontal position error, cm), and satellite ID.

3.2. Data Processing

For each GPS record, the first step of data processing is to change the geodetic coordinates (latitude and longitude) into geographical coordinates (X and Y coordinate) as well as the time and date in UTC value into local time. Being crucial measures for examining the quality of GPS records, satellite ID, EHPE, speed, and position (altitude, longitude, and latitude) are applied to remove invalid records in the dataset. Then GPS points are combined into trips and activities. Zong*et al.* [19] show the detailed process of identifying trips and activities with GPS data. Consisting of four sub-steps, namely, dividing status segments, identifying activities, recognizing trips, and determining intermediate stops, the identification process can determine trips and activities based on GPS logs with a high level of accuracy [19].

Eight typical identification parameters, as shown in Table 1, concerning the speed, acceleration, travel time, and trip distance of each trip, are then employed to represent the mode characteristics based on previous studies [7]. Then, the threshold ranges for each identification parameter are defined, according to the traffic condition and complex building environment in Beijing as well as the results that previous studies presented [20]. The thresholds of identification parameters (shown in Table 1) are then used to filter the data samples. After data filtering, the final sample consists of 85,120 GPS trips.

Table 1. The thresholds of identification parameters.

Identification Parameters	Threshold Ranges
Average speed (km/h)	1–110
Maximum speed (km/h)	1–110
75th percentile of speed (km/h)	0–110
Acceleration (km/h^2)	0–36,000
75th percentile of acceleration (km/h^2)	0–36,000
Travel time (h)	0–10
Trip distance (km)	0–1000
Standard deviation of speed	0–100

The daily activity and travel pattern of respondent #010 are shown in Figure 1. The major trip characteristics as well as the starting and ending time of each trip are also presented.

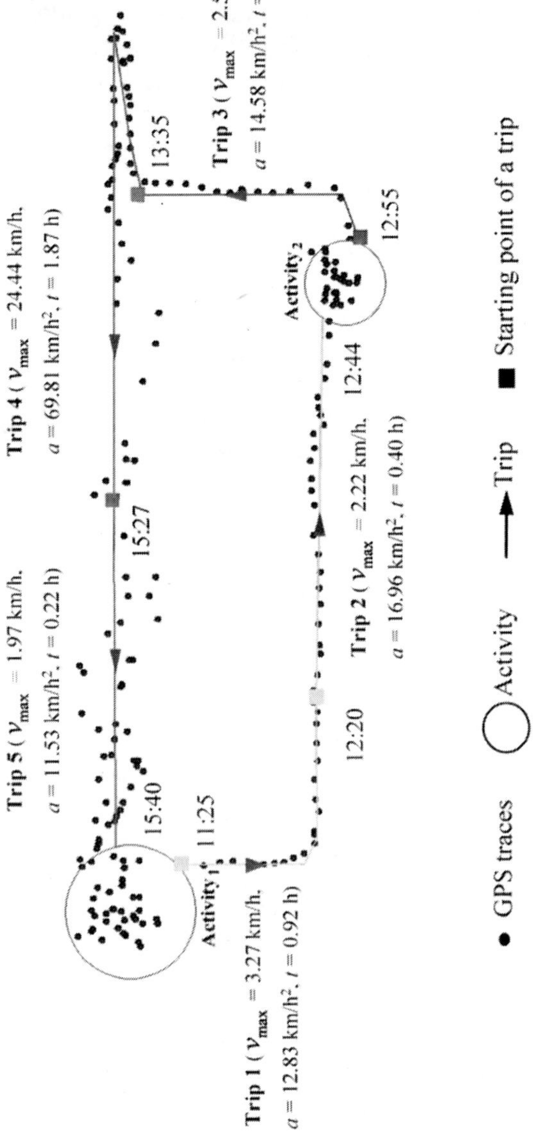

Figure 1. GPS traces and daily activity travel pattern of respondent #010.

4. METHODOLOGIES

In this section the major methods that will be used in mode detection are discussed. Being one type of SVM, SVC will be leveraged to identifying the traffic mode with GPS data. Compared to the multinomial logit (MNL) model, which has the limitation of determining alternatives with significant correlation (such as bus and car), SVC can provide us with higher prediction accuracy. Besides, GA will also be employed to optimize the major parameters in SVC in order to enhance the solution quality and calculation efficiency of the model.

4.1. SVC

Being a popular disaggregate model, the MNL model has been widely used in mode prediction and identification. However, it imposes the restriction that the distribution of the random error terms is independent and identical over alternatives. This restriction leads to limitation of random taste variation and the independence of irrelevant alternatives, which causes the cross-elastics among all pairs of alternatives to be identical [21]. Compared to the MNL model, SVC has been more frequently applied in modeling disaggregate choices in recent years due to its ability to enhance the prediction accuracy and calculation efficiency [22]. Therefore, this paper will introduce SVC in estimating of travel mode with GPS data.

SVC is one of the SVM (*i.e.*, machine learning) methods analyzing data and recognizing patterns. Given a set of input-output data pairs (*x*1, *y*1), (*x*2, *y*2), ..., (*xl*, *yl*) (xi∈X⊆Rm, yi∈Y⊆Rn, *l* is the number of training samples) that are randomly and independently generated from an unknown function, SVM estimates the function using the following Equation [23]:

$$f(x) = w \cdot \Phi(x) + b \qquad w, x \in R^m, b \in R^n, \tag{1}$$

where Φ(x) represents the high-dimensional feature spaces that are nonlinearly mapped from the input space *x*. *w* denotes a parameter vector and *b* is the threshold [24]. If the interpretation *y* only takes category values, *i.e.*, −1 and +1, it denotes SVC. Otherwise, if the domain of output space *y* contains continuous real values, the learning problem then refers to Support Vector Regression (SVR) [25].

For classification about the training data, SVM's linear soft margin algorithm is to solve the following regularized risk function:

$$MinJ = \frac{1}{2}\|w\|^2 + C \cdot R_{emp}[f] \tag{2}$$

The first term $\frac{1}{2}\|w\|^2$ is called the regularized term, which is used as a measurement of the function flatness. The second term Remp[f] is the so-called loss function to measure the empirical error. *C* is a regularization constant that determines the trade-off between the training error and the generalization performance. Here, the ε-insensitive loss function is employed to measure empirical error:

$$|y - f(x)|_\varepsilon = \max\{0, |y - f(x)| - \varepsilon\} \tag{3}$$

Equation (3) defines a ε tube (shown in Figure 2). The loss is zero if the predicted value is within the tube. If it is outside the tube, the loss is the magnitude of the difference between the predicted value and the radius *ε* of the tube. Both *C* and *ε*are user-determined parameters. Two positive slack variables ξ, ξ* are used to cope with the constraints of the optimization problem. To get the estimation of *w* and *b*, Equation (2) can be transformed to a primal objective function, Equation (4):

$$MinJ = \frac{1}{2}\|w\|^2 + C\sum_{i=1}^{s}(\xi_i^* + \xi_i)$$

$$s.t.\begin{cases} y_i - w\cdot\Phi(x_i) - b \le \varepsilon + \xi_i^* \\ w\cdot\Phi(x_i) + b - y_i \le \varepsilon + \xi_i \\ \xi_i^*, \xi_i \ge 0 \end{cases} \tag{4}$$

This constrained optimization problem is solved by using the following primal Lagrangian form:

$$L = \frac{1}{2}\|w\|^2 + C\sum_{i=1}^{s}(\xi_i^* + \xi_i) - \sum_{i=1}^{s}(\eta_i\xi_i + \eta_i^*\xi_i^*) - \sum_{i=1}^{s}\alpha_i(\varepsilon + \xi_i - y_i + w\cdot\Phi(x_i) + b)$$
$$-\sum_{i=1}^{s}\alpha_i^*(\varepsilon + \xi^* - y_i - w\cdot\Phi(x_i) + b) \tag{5}$$

where *L* is the Lagrangian, and ηi, η∗i, αi, α∗i are Lagrange multipliers. Hence the dual variables in Equation (5) have to satisfy the positive constraints:

$$\eta_i, \eta_i^*, \alpha_i, \alpha_i^* \geq 0 \tag{6}$$

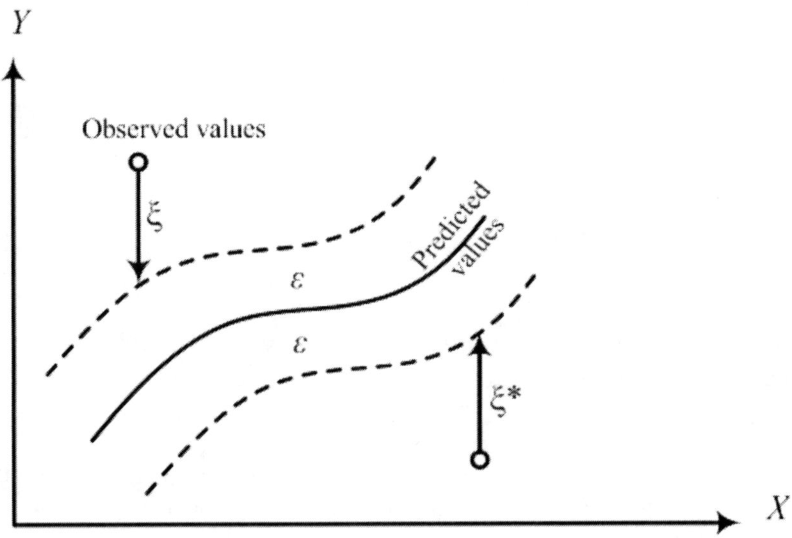

Figure 2. The parameters for SVC.

The above problem can be converted into a dual problem where the task is to optimize the Lagrangian multipliers, αi andα∗i. The dual problem contains a quadratic objective function of αi and α∗i with one linear constraint:

$$Max J = -\frac{1}{2}\sum_{i,j=1}^{s}(\alpha_i^* - \alpha_i)(\alpha_j^* - \alpha_j)(\Phi(x_i)\cdot\Phi(x_j)) + \sum_{i=1}^{s}\alpha_i^*(y_i - \varepsilon) - \sum_{i=1}^{s}\alpha_i(y_i + \varepsilon)$$

$$s.t.\begin{cases}\sum_{i=1}^{s}\alpha_i = \sum_{i=1}^{s}\alpha_i^* \\ 0 \leq \alpha_i \leq C \\ 0 \leq \alpha_i^* \leq C\end{cases} \tag{7}$$

Let

$$w - \sum_{i=1}^{s} (\alpha_i - \alpha_i^*) x_i = 0 \tag{8}$$

Thus,

$$f(x) = \sum_{i=1}^{s} (\alpha_i - \alpha_i^*) \Phi(x_i) \cdot \Phi(x_j) + b \tag{9}$$

By introducing kernel function *K*(*xi*,*xj*), Equation (9) can be rewritten as follows:

$$f(x) = \sum_{i=1}^{s} (\alpha_i - \alpha_i^*) K(x_i, x_j) + b \tag{10}$$

where *K*(*xi*, *xj*) is the so-called kernel function which is equal to the inner product of two vectors *xi* and *xj* in the feature space Φ(xi) and Φ(xj); that is, K(xi,xj)=Φ(xi)·Φ(xj).

The kernel function would be more simply without the compution of Φ(X). Some popular kernel functions are the linear kernel, polynomial kernel, and radial-basis function (RBF) kernel. Using different kernel functions, one can construct different learning machines with arbitrary types of decision surfaces. In general, the RBF kernel, as a nonlinearly kernel function, is a reasonable first choice [26]. Thus, the RBF kernel is chosen in this work:

$$K(x_i, x_j) = \exp\left(-\frac{\left\| x_i - x_j \right\|^2}{2\sigma^2}\right) \tag{11}$$

where σ is a parameter that determines the area of influence this support vector has over the data space.

As user-determined parameters in SVC and the RBF kernel, *C*, ε, and σ will greatly influence the estimation efficiency and prediction accuracy of the models, especially for large-scale or real-time feature practice application. This paper will optimize the parameters by employing GA. Being a part of evolutionary computing, GA is a rapidly growing area of artificial intelligence. The process of GA is presented as follows.

4.2. Parameter Optimization with GA

Being the key elements in SVC, the parameters *C*, ε, and σ directly decide the prediction performance of the model. Therefore, the parameter optimization is an important factor for improving the prediction accuracy of SVC. In this paper, GA is applied to optimize *C*, ε, and σ in SVC. GA is inspired by evolutionary biology processes like inheritance, selection, crossover, and mutation. Based on a fitness function, GA attempts to retain relatively good genetic information from generation to generation. The process of GA can be divided into six steps, which will be described briefly in the following section.

4.2.1. Encoding of Chromosomes

GA starts with a set of solutions (represented by chromosomes) called population. The individuals comprising the population are known as chromosomes. In most GA applications, the chromosomes are encoded as a series of zeroes and ones, or a binary bit string. For the mode detection model with SVC, the real encodings were adopted since the parameters*C*, ε, and σ are continuous-valued. To represent the parameters in SVC, each chromosome consists of genn1, genn2, and genn3(*n* refers to the current generation), which represent three parameters, respectively.

To reduce the search space referring to previous literature using SVC, the three parameters should within the range C∈[2−6,26], ε∈[2−12,2−1], and σ∈[0,2] [27]. An example of the encoding of a chromosome is shown in Table 2.

Table 2. An example of chromosome encoding.

$2^{-6} \leq C \leq 2^{6}$	$2^{-12} \leq \varepsilon \leq 2^{-1}$	$0 \leq \sigma \leq 2$
$gen^{n}{}_{1}$	$gen^{n}{}_{2}$	$gen^{n}{}_{3}$

4.2.2. Fitness Function

Fitness function determines possible solutions to the problem and is used to estimate the quality of the represented solution (chromosome). For parameter optimizations in SVC, the best solution is able to maximize the accuracy rate of prediction. Generally, GA is an optimal searching method to find the maximum fitness of the individual chromosome. Thus, Hit ratio is adopted in this paper. Here, Hit ratio (*HitR*) refers to the fitting degree of the identification results to the observed samples.

$$fit = \frac{N_1}{N_2} \tag{12}$$

where *N*1 is the number of hit records (travel mode) predicted by the model and *N*2 is the total number of observations.

4.2.3. Crossover Operator

Crossover is a reproduction technique that takes two parent chromosomes and produces two child chromosomes. In this paper, an arithmetic crossover is used to create new offspring [28].

$$\begin{aligned} gen^{n}_{k,I} &= \alpha_i gen^{n-1}_{k,I} + (1-\alpha_k) gen^{n-1}_{k,II} \\ gen^{n}_{k,II} &= \alpha_i gen^{n-1}_{k,II} + (1-\alpha_k) gen^{n-1}_{k,I} \end{aligned} \tag{13}$$

where genn−1k,I, genn−1k,II is a pair of "parent" chromosomes; gennk,I, gennk,II is a pair of "children" chromosomes; αk is a random number between (0, 1); and k∈[1,2,3] (*k* is the total genes of the crossover operation). Table 3 shows the parents selected for crossover. When k = 1 and αk = 0.4, the children chromosomes after crossover are shown in Table 4.

Table 3. An example of chromosome encoding.

Parent I	$gen^{n-1}{}_{1,I}$	$gen^{n-1}{}_{2,I}$	$gen^{n-1}{}_{3,I}$
Parent II	$gen^{n-1}{}_{1,II}$	$gen^{n-1}{}_{2,II}$	$gen^{n-1}{}_{3,II}$

Table 4. Children chromosomes after crossover.

Children I	$0.4gen^{n}{}_{1,I} + 0.6gen^{n}{}_{1,II}$	$gen^{n-1}{}_{2,I}$	$gen^{n-1}{}_{3,I}$
Children II	$0.4gen^{n}{}_{1,II} + 0.6gen^{n}{}_{1,I}$	$gen^{n-1}{}_{2,II}$	$gen^{n-1}{}_{3,II}$

4.2.4. Mutation Operator

Mutation is a common reproduction operator used for finding new points in the searching space to evaluate. A genetic mutation operation is used in this paper [27].

Assuming a chromosome is G=(genn1,genn2,genn3), if the genn1 is selected for the mutation, the mutation can be shown with Equation (14):

$$G' = (gen_1^{n-1}, gen_2^{n-1}, gen_3^{n-1})$$
$$gen_1^n = \begin{cases} gen_1^{n-1} + \Delta(n, gen_{1,\max}^n - gen_1^{n-1}) & if \quad random(0,1) = 0 \\ gen_1^{n-1} + \Delta(n, gen_1^{n-1} - gen_{1,\min}^n) & if \quad random(0,1) = 1 \end{cases} \tag{14}$$

Then, the function Δ(n,y) returns a value between [0, *y*] given in Equation (15):

$$\Delta(n, y) = y \times (1 - r^{(1-n/T_{\max})\lambda}) \tag{15}$$

where *r* is a random number between [0,1]; *T*max is maximum number of generations; and λ = 3. This property causes this operation to make a uniform search in the initial space when *n* is small, and a very local one in later stages.

To deal with the problem that the mutation may violate the parameters' constraints, this paper will assign a relatively high weight to reduce their probability of being selected in the following search [27,29].

4.2.5. Termination

The search continues until *HitRn-HitRn-1* < 0.001% or the number of generation reaches the maximum number of generations *T*max, which is set to be 5000 [30].

4.2.6. The procedure of GA

The major steps of GA are shown in Figure 3.

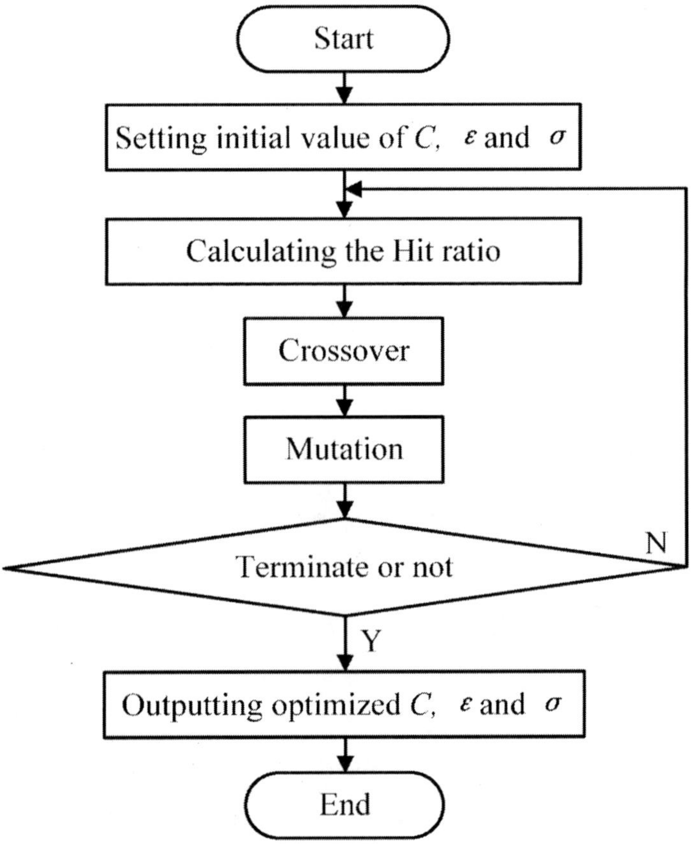

Figure 3. The basic procedure of GA.

5. IDENTIFICATION MODEL

In this section the detailed setting of parameters in SVC and GA as well as the modeling procedure of mode detection are introduced. The models are estimated with the survey data in Beijing. Compared to the observed records provided by the survey data, success rates of 100%, 100%, 88.9%, 92.7%, and 80.0% were obtained for the detection of walking, bicycle, subway, bus, and car, respectively. This indicates that the developed model shows a high level of accuracy rate for mode identification. For a case study, the identified and observed mode choices of respondent #010 are presented.

5.1. Alternatives

According to the travel survey data, the real mode share in Beijing in 2010 was calculated. The results are shown in Table 5. The results reveal that car and bus are the two most popular travel modes in Beijing. For mode identification using GPS data, buses and cars have great correlation due to their similar characteristics, especially concerning the speed and acceleration in almost the same traffic conditions. Therefore, as mentioned above, SVC, instead of the MNL model, will be employed to establish the mode identification model in this paper [31,32].

Table 5. Mode share in Beijing, 2010.

Mode	Walking	Bicycle	Subway	Bus	Car	Other
Percentage (%)	10.7	15	6.9	22.2	30	15.2

5.2. Variables

Based on a preliminary correlation test, eight variables related to the characteristics of travel modes were selected, as shown in Table 6.

Table 6. Variables in the mode identification model.

Variables		Mean [1]	S.D. [2]	Variables		Mean	S.D.
Average Speed (km/h)	v_{walk}	3.00	2.49	*Average acceleration* (km/h²)	a_{walk}	57.97	43.26
	$v_{bicycle}$	8.47	8.20		$a_{bicycle}$	60.90	80.81
	v_{subway}	11.21	13.32		a_{subway}	73.47	118.94
	v_{bus}	13.27	14.21		a_{bus}	78.99	115.95
	v_{car}	14.54	9.20		a_{car}	84.58	103.59
Maximum speed (km/h)	$v_{max\text{-}walk}$	15.35	7.45	*75th percentile of acceleration* (km/h²)	$a_{75th\%\text{-}walk}$	434.20	400.86
	$v_{max\text{-}bicycle}$	20.87	14.73		$a_{75th\%\text{-}bicycle}$	864.85	431.41
	$v_{max\text{-}subway}$	44.51	30.22		$a_{75th\%\text{-}subway}$	722.90	754.76
	$v_{max\text{-}bus}$	50.23	19.33		$a_{75th\%\text{-}bus}$	2012.55	1203.71
	$v_{max\text{-}car}$	53.25	18.73		$a_{75th\%\text{-}car}$	2290.54	1327.73

Variables		Mean [1]	S.D. [2]	Variables		Mean	S.D.
75th percentile of speed (km/h)	$v_{75th\%\text{-}walk}$	3.62	2.01	*Travel time* (h)	t_{walk}	2.56	3.17
	$v_{75th\%\text{-}bicycle}$	12.44	10.47		$t_{bicycle}$	3.46	4.92
	$v_{75th\%\text{-}subway}$	24.19	31.92		t_{subway}	0.56	0.30
	$v_{75th\%\text{-}bus}$	25.15	12.32		t_{bus}	1.21	1.12
	$v_{75th\%\text{-}car}$	24.48	15.47		t_{car}	0.64	0.73
Standard deviation of speed	$\sigma_{walking}$	4.80	7.68	*Trip distance* (km)	d_{walk}	0.36	0.29
	$\sigma_{bicycle}$	0.92	0.52		$d_{bicycle}$	2.03	2.76
	σ_{subway}	2.53	1.80		d_{subway}	4.52	4.96
	σ_{bus}	2.37	2.54		d_{bus}	7.59	9.26
	σ_{car}	1.62	3.44		d_{car}	8.17	7.48

Note: [1] Mean denotes mean value; [2] S.D. refers to standard deviation.

5.3. Estimation Results

There are three major GA parameters, namely *pc*, *pm*, and *psize*. In general, *pc* varies from 0.3 to 0.9 and *pm* varies from 0.01 to 0.1, while *psize* is the population size, which is set according to the size of the samples. Considering the features of mode identification, the condition of survey data, and the previous studies [27,28,33,34] related to GA, *pc*, *pm*, and *psize* are set at 0.6, 0.06, and 80, respectively. The convergence of the calculation by GA indicates that the Hit ratio increases slowly after the 3000th generation. The highest Hit ratio appears in about the 3500th generation, and remains almost unchanged after that. The three parameters, *i.e.*, C, ε, and σ were then optimized as 84.54, 0.0009, and 0.7367, respectively, with the best optimization value among the 10 results for the practical mode detection model.

The estimation results of the SVC model are shown in Table 7. The results indicate that the factors related to acceleration, travel speed, and travel time are the major ones that should be considered in travel mode detection, while that regarding trip distance does not have significant impact on mode decision. One of the reasons for this is that there is correlation between travel time and trip distance. In detecting walking trips, travel time and maximum speed are important factors—that is, the longer a trip takes or the larger its maximum speed is, there is less probability that its mode is walking. The results also show that there are three major factors, *i.e.*, the 75th percentile of speed, maximum speed, and acceleration, which impact the detection of a bicycle. In detail, a trip with low speed and acceleration tends to be by bicycle. On the contrary, a trip with high speed and acceleration is more likely to be on the subway. The determination of a bus is similar to that of subway, except that the coefficient of maximum speed and acceleration is smaller than that of the subway. This reveals that the higher a trip's speed and acceleration, the more probability that it refers to the mode of subway.

5.4. Mode Identification

The travel mode of each trip in the sample data is determined by using the developed model. Figure 4 shows respondent #010's daily mode choices identified by the model

Table 7. Estimation results of the mode detection model.

Variables	Coefficient	Standard Error	T-statistic
$t_{walking}$	−6.009	3.375	−1.78
$v_{max\text{-}walking}$	−0.861	0.394	−2.19
$v_{75th\%\text{-}bicycle}$	−0.351	0.172	−2.04
$a_{bicycle}$	−0.281	0.120	−2.34
$v_{max\text{-}bicycle}$	−0.412	0.157	−2.62
$a_{75th\%\text{-}subway}$	−0.004	0.001	−4.00
a_{subway}	0.172	0.049	3.51
$v_{max\text{-}subway}$	0.474	0.169	2.80
$v_{max\text{-}bus}$	0.006	0.002	3.00
a_{bus}	0.061	0.015	4.07

Note: Dependent variable = mode; Number of sample = 85,120.

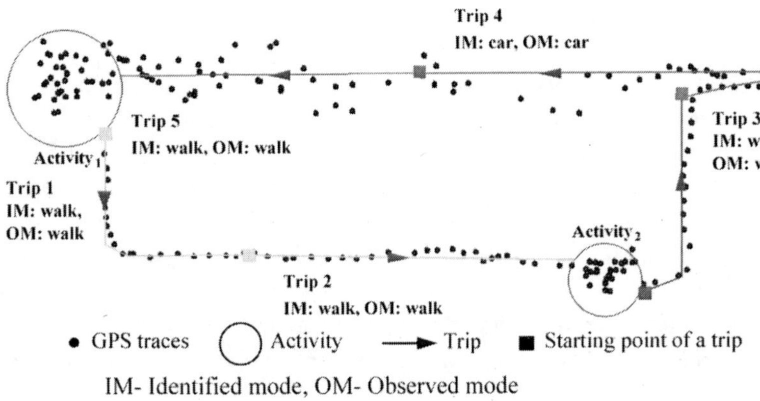

Figure 4. Identified and observed mode choices of respondent #010.

With the developed model, the mode choices of all the trips in the GPS data are determined. By comparing the predicted results and the observed records, the Hit ratios of the model are shown in Table 8. The estimation results show high accuracy for detection of walking, bicycle, subway, and bus, and that of car is also acceptable. Further analysis reveals that the reason for the relatively lower accuracy of car and subway determination is that the model sometimes makes mistakes in distinguishing between car and subway. Furthermore, compared to the detection results presented by Gong *et al.* [3]—*i.e.*, 92.4%, 65.5%, 62.5%, and 84.1% for walking, subway, bus, and car, respectively—the detection results of this work are better. This also represented the good performance of the SVC model in mode detection.

6. CONCLUSIONS

In this paper, an SVC model was constructed for mode detection with GPS survey data. GA was used in optimizing the parameters in SVC. The results indicate that the developed model shows a high level of accuracy for mode identification. Our findings can significantly enhance the efficiency and accuracy of travel survey and data processing. They also serve as a foundation for a future model system of full-scale travel information identification with GPS data. Moreover, by providing crucial travel information, the results contribute to the modeling and analyzing of travel behavior and are readily applicable to a wide range of transportation practice.

The estimation results also reveal that further study should be conducted with respect to high-accuracy detection of subway, bus, and car. One of the potential methods is to distinguish subway and bus from car based on the GIS information. Therefore, a future study could make potential progress by combining GPS data with GIS technology to determine travel modes. Moreover, although the sample size is relatively large compared to some of previous studies, it was not representative enough for the general population. Further study should pay more attention to the process of sampling and data collection to eliminate sampling bias.

Table 8. Verification results of the mode detection model.

	Identified by algorithm as					Total trip	Correct	Success rate
	Walking	Bicycle	Subway	Bus	Car	segments	segments	(%)
Walking	24,961	0	0	0	0	24,961	24,961	100.0
Bicycle	0	12,685	0	0	0	12,685	12,685	100.0
Subway	0	0	3250	204	202	3656	3250	88.9
Bus	206	409	0	18,209	818	19,642	18,209	92.7
Car	129	205	2864	1637	19,341	24,176	19,341	80.0
Total	25,296	13,299	6114	20,050	20,361	85,120	78,446	92.2

ACKNOWLEDGMENTS

The research was funded by the National Natural Science Foundation of China (50908099), the Humanity and Social Science Youth Foundation of the Ministry of Education (14YJC630225), the China Postdoctoral Science Special Foundation (2014M551191), and Jilin University's Outstanding Youth fund (2013JQ007). Special thanks to the anonymous reviewers whose valuable comments helped tremendously in improving this paper.

AUTHOR CONTRIBUTIONS

The research scheme was mainly designed by Fang Zong and Yu Bai. Xiao Wang, Yixin Yuan and Yanan He performed the research and analyzed the data. The paper was mainly written by Fang Zong. All authors have read and approved the final manuscript.

CONFLITS OF INTEREST

The authors declare no conflict of interest.

REFERENCES

1. Stopher, P.; Greaves, S.P. Household travel surveys: Where are we going? *Transport Res A.* **2007**, *41*, 367–381.

2. Stopher, P.R.; Metcalf, H.M.A. *Metcalf, Synthesis of Highway Practice No. 236: Methods for Household Travel Surveys*; NA Press: Washington, DC, USA, 1996.

3. Gong, H.; Chen, C.; Bialostozky, E.; Lawson, C.T. A GPS/GIS method for travel mode detection in New York City.*Comput. Environ. Urban Syst.* **2012**, *36*, 131–139.

4. Liu, J.; Zheng, H.J. Post-processing procedures for passive GPS based travel survey. *Procedia Soc. Behav. Sci.* **2013**, *96*, 310–319.

5. Bohte, W.; Maat, K. Deriving and validating trip destinations and modes for multi-day GPS based travel surveys: An application in the Netherlands. *Transp. Res. C* **2009**, *17*, 285–297.

6. Byon, Y.J.; Abdulhai, B.; Shalaby, A.S. Impact of sampling rate of GPS-enabled cell phones on mode detectionand GIS map matching performance. In Proceedings of the Transportation Research Board 86th Annual Meeting, Washington, DC, USA, 21–25 January 2007.

7. Gonzalez, P.; Weinstein, J.; Barbeau, S. Automating mode detection using neural networks and assisted GPS data collected using GPS-enabled mobile phones. In Proceedings of the 15th World Congress on ITS, New York, NY, USA, 10–15 January 2008; pp. 297–300.

8. Patterson, D.J.; Liao, L.; Fox, D.; Kautz, H. Inferring high-level behavior from low-level sensors. In *UbiComp 2003: Ubiquitous Computing*; Anind, K.D., Albrecht, S., Joseph, F.M., Eds.; Springer: Berlin/Heidelberg, Germany, 2003; pp. 73–89.

9. Zheng, Y.; Liu, L.; Wang, L.; Xie, X. Learning transportation mode from raw GPS data for geographic applications on the web. In Proceedings of the WWW 2008, Beijing, China, 21–25 April 2008; pp. 247–256.

10. Reddy, S.; Burke, J.; Estrin, D.; Hansen, M.; Srivastava, M. Determining transportation mode on mobile phones. In Proceedings of the 12th IEEE International Symposium on Wearable Computers, Pittsburgh, PA, USA, 28 September–1 October 2008; pp. 25–28.

11. Moiseeva, A.; Timmermans, H. Imputing relevant information from multi-day GPS tracers for retail planning and management using data fusion and context-sensitive learning. *J. Retail. Consum. Serv.* **2010**, *17*, 189–199.

12. Feng, T.; Timmermans, H.J.P. Transportation mode recognition using GPS and accelerometer data. *Transp. Res. C* **2013**,*37*, 118–130.

13. Zhang, L.; Dalyot, S.; Eggert, D.; Sester, M. Multi-stage approach to travel-mode segmentation and classification of GPS traces. In Proceedings of the ISPRS Guilin 2011 Workshop on International Archives of the Photogrammetry, Remote Sensing and Spatial Information Sciences, Guilin, China, 20–21 October 2011; pp. 87–93.

14. Pereira, F.; Carrion, C.; Zhao, F.; Cottrill, C.D.; Zegras, C.; Ben-Akiva, M. The future mobility survey: Overview and preliminary evaluation. In Proceedings of the 10th International Conference of Eastern Asia Society for Transportation Studies, Taipei, Taiwan, 9–12 September 2013; p. 31.

15. Zhang, Z.H. Deriving Trip Information from GPS Trajectories. Ph.D. Thesis, East China Normal University, Shanghai, China, June 2010. (In Chinese).

16. Vij, A.; Carrel, A.; Walker, J.L. Incorporating the influence of latent modal preferences on travel mode choice behavior.*Transp. Res. A* **2013**, *54*, 164–178.

17. Du, J.; Aultman-Hall, L. Increasing the accuracy of trip rate information from passive multi-day GPS travel datasets: Automatic trip end identification issues. *Transp. Res. A* **2007**, *41*, 220–232.

18. Bolbol, A.; Cheng, T.; Tsapakis, I. A spatio-temporal approach for identifying the sample size for transport mode detection from GPS-based travel surveys: A case study of London's road network. *Transp. Res. C* **2014**, *43*, 176–187.

19. Zong, F.; Wang, X.; Zhang, H.Y.; Bai, Y. Identifying trip-activity-intermediate stop using GPS-based travel survey data.*J. SCUT. (Nat. Sci. Ed.)* **2015**, *43*, 28–32. (In Chinese).

20. Sehonfelder, S.; Samaga, U. Where do You Want to Go Today?—More Observations on Daily Mobility. In Proceedings of the 3rd Swiss Transport Research Conference (STRC), Ascona, Switzerland, 19–21 March 2003; Available online: http://www.strc.ch/Paper/Schoen.pdf (accessed on 28 November 2004).

21. Wen, C.H.; Koppelman, F.S. The generalized nested logit model. *Transp. Res. B* **2001**, *35*, 627–641.

22. Zong, F.; Jia, H.F.; Pan, X.; Wu, Y. Prediction of commuter's daily time allocation. *PROMET Traffic Transp.* **2013**, *25*, 445–455.

23. Sung, K.; Poggio, T. Example-based learning for view-based human face detection. *IEEE Trans. Pattern Anal. Mach. Intell.* **1998**, *20*, 39–50.

24. Vapnik, V. *The Nature of Statistical Learning Theory*; Springer: New York, NY, USA, 1995.

25. Li, Y.M.; Gong, S.G.; Liddell, H.M. Support Vector Regression and Classification based multi-view face detection and recognition. In Proceedings of the IEEE International Conference on Automatic Face and Gesture Recognition, Grenoble, France, 28–30 March 2000; pp. 300–305.

26. Dong, B.; Cao, C.; Lee, S.E. Applying Support Vector Machines to predict building energy consumption in tropical region. *Energ. Build.* **2005**, *37*, 545–553.

27. Yao, J.B.; Yao, B.Z.; Li, L.; Jiang, Y.L. Hybrid model for displacement prediction of tunnel surrounding rock. *Neural Netw. World* **2012**, *22*, 263–275.

28. Yu, B.; Yang, Z.Z.; Cheng, C. Optimizing the distribution of shopping centers with parallel genetic algorithm. *Eng. Appl. Artif. Intell.* **2007**, *20*, 215–223.

29. Yao, B.Z.; Hu, P.; Lu, X.H.; Gao, J.J.; Zhang, M.H. Transit network design based on travel time reliability. *Transp. Res. C***2014**, *43*, 233–248.

30. Zong, F.; Lin, H.Y.; Yu, B.; Pan, X. Daily commute time prediction based on genetic algorithm. *Math. Probl. Eng.* **2012**.

31. Este, A.; Gringoli, F.; Salgarelli, L. Support Vector Machines for TCP traffic classification. *Comput. Netw.* **2009**, *53*, 2476–2490.

32. Anguita, D.; Boni, A.; Ridella, S. Evaluating the generalization ability of Support Vector Machines through the bootstrap. *Neural Process. Lett.* **2000**, *11*, 51–58.

33. Wang, Y.Q.; Li, Y.; Wang, Q.; Lv, Y.L.; Wang, S.Y.; Chen, X.; Yu, X.X.; Jiang, W.; Li, X. Computational identification of human long intergenic non-coding RNAs using a GA-SVM algorithm. *Gene* **2014**, *533*, 94–99.

34. Lee, Y.; Lee, J. Binary tree optimization using genetic algorithm for multiclass support vector machine. *Expert Syst. Appl.* **2015**, *42*, 3843–3851.

Index